Cast Off your Bow Lines

Nate,
You have been such a part of
this journey. Thank you

Cap'm John

John V. Clarke

CAST OFF YOUR BOW LINES
Copyright © 2021 by John V. Clarke

All rights reserved. Neither this publication nor any part of this publication may be reproduced or transmitted in any form or by any means, electronic or mechanical, including photocopying, recording or any information storage and retrieval system, without permission in writing from the author.

Unless otherwise indicated, all scripture taken from the Holy Bible, American Standard Version, which is in the public domain. Scripture marked (NKJV) taken from the New King James Version®. Copyright © 1982 by Thomas Nelson. Used by permission. All rights reserved.

Printed in Canada

Print ISBN: 978-1-4866-2121-7
eBook ISBN: 978-1-4866-2122-4

Word Alive Press
119 De Baets Street, Winnipeg, MB R2J 3R9
www.wordalivepress.ca

Cataloguing in Publication may be obtained through Library and Archives Canada

Dedication

I dedicate this book to my beloved wife of fifty years. That's only fitting considering that she has not only been my sailing buddy and lover all this time but also my secret weapon in the writing of books. This is now the fourth manuscript in which she has followed behind my somewhat rough storytelling to bring a smoother flow of content, refinement of verbiage, and, of course, grammar correction to my more cavalier beginnings. At the same time, she has respected my voice and left the writing style uniquely mine. This she has done throughout our life together conceiving and realizing our dreams—following my lead yet lending her own unique personhood and style to the mix of every endeavour. If the truth be known, it was actually her enthusiasm and encouragement that were the primary catalysts to get us to venture out to cross the ocean. I am forever grateful for a life partner who gets me and my pioneer spirit and also has the pluck to throw off her own bow lines to join in the adventure. This book is not only dedicated to her but remains uniquely hers.

I love you, my darlin' … Captain John

Contents

Dedication	iii
About the Author	vii
Endorsement	ix
Preface	xi

Book One: Throw off the Bow Lines

1.	Cast Upon the Waters	1
2.	An Affair	8
3.	An Unlikely Prophet	13

Book Two: Sail Away from Your Safe Harbour

4.	Bernie	28
5.	Miyagi in the Belly of the Whale	36
6.	Anchoring Can Be a Drag	46
7.	The Charter from Hell	54
8.	The Call of Mother Ocean	79

Book Three: Catch the Trade Winds in Your Sails

9.	Bon Voyage	89
10.	Juan de Puca	94
11.	Boggled	100
12.	There Be Treasure	109
13.	Smilin' Jack	115
14.	Catch the Trade Winds	123
15.	Fiddlin'	129
16.	Hilo Heaven	141

17.	Captain Bluewater	154
18.	Cetacean Visitations	165

Book Four: Explore ... Dream ... Discover

19.	Legacy	172
20.	Princess Encounters	184
	Another Book by John Clarke	211

About the Author

John Clarke was born on September 10, 1950, in Seattle, Washington. He grew up on his parents' lakeside property on Mercer Island with his four brothers and one sister. He graduated with a teaching degree from Western Washington State University in Bellingham, where he was impacted by the charismatic Christian movement of the day. In 1970, John married his teenage sweetheart, Lyza, and moved to White Rock, British Columbia, in Canada, in order to work with youth. There they started an outreach coffeehouse on the White Rock beach and eventually led their following of young people into a local nondenominational congregation.

At that time, John's vocation took a turn from teaching to pastoring. For the next thirty-five years, John and Lyza raised their family of four children and served as ministers at White Rock Christian Fellowship. After following this vocational calling for such an extended period, their life took another radical turn when they became coastal skippers operating a charter business called Pacific Encounters on a large classic sailboat. This late-in-life transition is the subject of this, his latest book, inspired by the theme of daring to go outside one's comfort zones and embracing the adventures life still has in store.

John and Lyza still live in White Rock by the sea, spending winters in their beachside condo and summers on their sailboat, *Porpoise*, running sailing charters with people from all over the globe. Their four children and seven grandchildren also join them regularly to cruise the pristine wilderness of the Inside Passage.

Endorsement

Two years ago, a couple, Erin and Adam, came aboard *Porpoise* as charter guests for a week. It turned out to be a serendipitous connection. As we got to know each other, we discovered many synergies between us, not least of which was that Erin was a PhD candidate focusing on Canadian literature at McMaster University. At the time of their cruise, I was finishing up the writing of this book. When I learned that she did professional editing on the side, I offered to send her a copy of the manuscript to get her opinion on it. The following is an excerpt from the notes she wrote to me about her impressions.

***

A charming and enjoyable read. There's a fluidity to the prose and an earnestness to the storytelling that's very appealing and totally captures the experience of being on *Porpoise* with you. The interventions back into childhood/family history are compelling, as they set a broad stage for the more technical details of sailing adventures.

I particularly like/am impressed by the balance you strike between human stories and technical information and knowledge, and I imagine it will appeal to a broad readership. There's a deftness to the narrative as it moves from the personal to the marine and back—lovely, informative, and engaging. I think particularly of the explanation of a "sea run cut," or how you artfully touch on what "hull speed' is while explaining how it made you feel to be at *Porpoise*'s helm for the first time.

The details of both your feelings and the spaces and experiences you encounter make for good, engaging reading, and the magic infused through the whole narrative—from Lyza's quote to your family's long-time love of Princess Louisa, and how you share that with charter clients (who have their own magic)—is absolutely lovely.

Cast Off Your Bow Lines

I love the inclusion of your mother's journal. It's an important addition to the text—women's voices from this time period are so often hidden and lost historically. Privileging her journal in your own book both honours her and helps her history be recorded. That's pretty special.

—Erin Ramlo
PhD Candidate
Department of English and Cultural Studies
McMaster University

Preface

Why should you read this book? If you've ever entertained fantasies of sailing away from it all, or even contemplated buying or renting a boat and learning to cruise, our story might motivate you to go and do it. If you sometimes feel "a wild call and a clear call that may not be denied," you can at least draw encouragement from our experiences, even if only from an armchair next to the fire. I suppose you could also read this book if you need some courage to step out beyond the confines of the pathway you've already worn out in your life. If you think there may be more to your story and it might be time to write a few brand-new chapters, may this true account give you inspiration.

Writing another book at this point in my life was just such an exercise. Acquiring a sailing vessel, starting a charter business on the British Columbia coast, and sailing across the Pacific Ocean in my late fifties was another. My wife and I both come from boating backgrounds and have sailed British Columbia coastal waters together for over fifty years now. With Pacific Encounters Sailing Adventures we get to share the fun with others from all over the world. So far, we have taken hundreds of people over the years onboard *Porpoise* each summer to cruise the coast for a week at a time. By means of the pen, we can vicariously take hundreds more—not only to the incomparable Inside Passage but also through our personal journey to get there.

That journey was unique. A providential quote attributed to a famous author became the unlikely catalyst for launching us into this late-in-life undertaking. What happened to us could happen to you. Just when you think your life has rounded the bell curve and is on the downhill, you might be surprised by joy. I hope you can get the salty taste of the sea and the fresh excitement of our odyssey by coming with us in the following pages.

Book One
Throw off the Bow Lines

Sea Fever

I must go down to the seas again, to the lonely sea and the sky,
And all I ask is a tall ship and a star to steer her by;
And the wheel's kick and the wind's song and the white sail's shaking,
And a grey mist on the sea's face, and a grey dawn breaking.
I must go down to the seas again, for the call of the running tide
Is a wild call and a clear call that may not be denied;
And all I ask is a windy day with the white clouds flying,
And the flung spray and the blown spume, and the sea-gulls crying.
I must go down to the seas again, to the vagrant gypsy life,
To the gull's way and the whale's way where the wind's like a whetted knife;
And all I ask is a merry yarn from a laughing fellow-rover,
And quiet sleep and a sweet dream when the long trick's over.

By John Masefield

1. Cast Upon the Waters

Plodding down the uneven slope, through tangles of salal and salmonberry bushes, my spirits were as dampened as the moss around me, shut out from sunlight under the high canopy of alder leaf. My current mood was in stark contrast to the sunny expectations that had drawn me up the ascent of this same non-existent trail earlier that morning. Then, it was all the thrill of discovery and anticipation of the unknown. Then, I'd been looking forward to my tramp homeward with a string of slimy fish strung through the gills and tied through the belt loop of my pants—my signature proof of wilderness prowess. Then, I imagined, I would victoriously mount the boarding ladder of the family motor sailer, *Nor'wester*, holding aloft my prizes to the inevitable exclamations from my parents and, hopefully, my brothers. Mother would declare, with her characteristic enthusiasm, "Oohhh, look at those beautiful fish Johnny caught. They're just wonderful, honey!" The encore would come from my father: "How 'bout that! Big J is the great hunter!"

Such hoopla was always an embarrassment I feigned I could do without, but in reality, I lived for it. Continued praise and notoriety would come for me at the dinner table, where the fish would be prominently featured. Now, though, instead of that whole imagined scenario, I was skulking back down the trail with one measly, somewhat mangled, minnow of a fish—hardly sufficient to do me proud. Back in the early morning hours, I'd been full of great expectations. Nearing the marge of the small lake, I met with the ethereal magic that such woodland pools inspire: wreaths of mist drifted low over greenish brown depths, and ducks and loons glided over glassy emerald green without a ripple, as though suspended in air. My ascent had felt akin to a pilgrimage, and I a devoted seeker on a holy quest.

Fishing was my way of extracting meaning and reward from such quests. I had long been initiated to the pleasure and anticipation of the first cast, made far out upon translucent waters. I knew that underneath some sunken log my quarry

Cast Off Your Bow Lines

awaited. First, the whipping sound of the rod through the air, the whirring clip of leader leaving the spool, and then the soft little plash of the lure landing over a promising spot. A moment of descent and then the sparkle of reflected silver or gold as the fluttering spinner brought a slight resistance to the returning line. I would watch, mesmerized, as the underwater butterfly danced seductively only a foot or so under the surface. The tension was always exquisite, because try as I might, I never saw them before the strike. Suddenly, the lure would stop in mid-flight and be carried off sideways like a halfback tackled midfield by a bloodthirsty linebacker. Only after this would the rod bend and jerk, the line spinning off the reel with the drag loosened. I knew then that I'd connected with another world.

The peaceful lake spell would be broken by the eruption of the sudden frantic contact between man and fish. Immediately my tentative perch on whatever log I was balanced would be in jeopardy as the fish broke out of his watery domain to furiously try to shake the hook. The spinner would make a tinkling sound as he repeatedly launched into the air. The intense skirmish would usually be a brief melee of fisherman and fish slipping and struggling to maintain their tenuous hold on their respective domains. Inevitably, one of them would lose the battle. After several grabs at the slippery prize, I would either secure him with a throttling grip on his gills or he would slip through my hands like a bar of soap in the shower to shoot back into the depths from whence he came—leaving me to nurse my disappointment. After the wrestling match, the stillness would return as though nothing had happened. I would fumble with my gear in order to repeat—with the intensity of an addict—the rituals of the quest again and again.

That morning, after casting for some time from logs resting on the perimeter of the lake, I eventually found an abandoned rowboat with crackling greyed timbers and covered with moldy lichen. Scrambling aboard, my runners were soon soaked by the water that seeped into the leaky craft and sloshed about my feet. I rowed slowly up and down the shore, towing my spinner and casting in every corner of the lake—anticipating a strike at any moment. Instead, the morning stillness slipped away, and the sun rose to its zenith, producing a buggy, itchy afternoon heat. I had nothing to show for all my efforts except one scrawny cutthroat dried out in the sun on the dinghy's back seat.

I finally gave up and headed despondently down the trail, with nothing to brag about to my big brother, who was to pick me up in the runabout later in the afternoon. When I arrived among the grassy reeds and logs piled along the rocky shore, neither he nor any boat was in sight, but at least it was cool under

Cast Upon the Waters

the shelter of large maples above me. I watched the little wavelets marking the final high tide of the day. The moon had been large on the previous nights, and the water would come much higher on the rock faces than in the days before.

The shore of this bay was like most other tidal areas I had frequented since childhood. One knows what and what not to expect after a while—logs, rocks, seaweed, broken shells, a few beach crabs. Nothing exceptional would be there to impress me after years of adolescent exploration of many such little coves. The water was clear and shallow for at least a hundred yards out from the edge, and then it dropped suddenly to about twenty feet deep, where it was a great place to catch Dungeness crab. Out farther there was great cod fishing off the kelp beds in the channels beyond.

From where I waited on the beach logs, I could see down through the water to a sandy mud bottom pockmarked with shrimp holes interspersed with weedy rocks, and I knew that seabed hosted a world of small creatures living out their daily adventures. I entertained myself by crushing a mussel and tossing it down in their midst, where I knew it would stir up trouble. For a minute or two it rested lightly on the sand. Then the mud and weeds began to stir. Little feelers and heads popped out from every nook and cranny. Within five minutes, I had incited a riot. Beach crabs, hermit crabs, and small bullheads dodging behind stones came sneaking and creeping magnetically toward the prize. The beach crabs were in the centre of the melee, tussling and tearing at the meat, surrounded by an assortment of small bullheads and hermit crabs scurrying around the perimeter and brawling over the scraps. No protein went to waste in that hungry environment. It was strange to view the drama in this watery diorama, knowing that by the morning's low tide this world would be gone, reduced to a lifeless flat of mud and rock. Little did I suspect, looking up from this distraction to the bay surrounding me, that much more magnificent creatures also foraged in this tidal domain.

I'm not sure what prompted me to start casting about in those shallow waters that late afternoon. It was most likely boredom. I decided to toss my spinner into this tidal world even though I'd been casting about all day. There was nothing else going on, and a fourteen-year-old needs to be doing something. Once, twice, my lure sailed out over the water. Suddenly, to my astonishment, my rod stiffened on the third cast. Thirty feet out, my spinner flew sideways, and my pole bent hard.

"Whaaaaaat? A fish?" Too big for a bullhead. Maybe a big sea perch? I focused on the spinner veering violently off to the left, and then the water erupted

3

Cast Off Your Bow Lines

in a splash and flash of silver. Big silver. The only fish to look and behave this way would be a salmon. Too excited to think, I yanked on the pole and fought the fish, staring intently into the water. I could clearly see a dark shape hurrying back and forth. Beside it, two more shadows appeared, chasing the spinner hooked in its mouth. They romped about in the shallows like dogs competing for a ball in a back yard. All three were huge. Frantically, I cranked on the reel, hauling in the line much too quickly as I precipitously dragged the flailing fish toward my log. It was a wonder he stayed on the line. There was no careful playing of this prize; I wanted him secured *now*! After a few aborted attempts of slashing at him with my tiny trout net and slipping off my perch into the water up to my knees, I finally made a hasty scoop at his head as he thrashed about the log. Miraculously, I snared him headfirst into the net and fell abruptly back against the log, holding my writhing treasure aloft and whooping and hollering in amazement. He was almost two feet long, with a green speckled back and silver sides. Along his throat were the characteristic orange red markings of a cutthroat trout. Unlike the emaciated lake trout still hanging from my belt, he was fat and flourishing, a feast to the eyes and a tasty treat for the palate. As I digested the import of this momentous catch and the impending glory that would soon fall upon me, I heard the sound of an outboard motor and turned to see my brother rounding the distant point on his way to pick me up in the small tender.

The ensuing commotion back on the family boat lived up to my highest expectations. I was lauded a hero and a discoverer and then honoured at the subsequent dinner that night, where the succulent orange flesh of my catch steamed on a platter in the centre of the table. I was further gratified the next day when my brothers, with poles in hand, rushed to my discovery spot on the log to see if they could match my catch. They had some success in subsequent years but were never quite able to achieve my prowess at this particular fishing hole. I became known by my Uncle George (who was a notoriously good fisherman) as "my nephew John, who discovered the big trout in that henceforth legendary bay."

Later in life I learned the secret of sea run cuts, as they are called—trout that swim down from lake to stream to tidal waters in search of more abundant food sources, growing big and tasty on a saltwater smorgasbord of shrimp, crabs, and smaller minnows that inhabit such flats. As time went on, I searched and found them in other lonely places up and down the British Columbia coast. But this spot has always remained the most magical and has consistently delivered up fish for the last fifty years. I dare say that most boaters and cabin dwellers who stay in that bay remain ignorant of these prize trout—indeed, the existence and

Cast Upon the Waters

location of these riches is a closely kept family secret to this day. One must keep buried treasure buried, except for making a surreptitious withdrawal from the trove once in a while.

Later in life, the natural revelations of that day pointed to life lessons in other realms of endeavour. A cutthroat trout discovered unexpectedly amid the commonplace woods and waters of the natural environment is a treasure indeed. With its wondrous greenish-gold skin, all speckled with silver and black spots and marked with the signature orange slash about its throat, it's as lovely as the most beautiful gem. Add to the glitter of its appearance the grace of its symmetrical form and the agility of its movements and you have beauty manifestly defined. Yet it remains hidden as it glides about its world until revealed in a sudden, startling display of leopard-like beauty leaping into the air to take your breath away. Though perishable, it is the solid gold of the natural world, like so many of its fellow creatures. Precious discoveries like these fish or like gold, silver, and precious stones, aren't found in the common alleyways of the world. They're tucked away in the remote recesses of creation: in the heart of a mountain, in the bed of a wild stream, on an uncharted shore, or in the depths of the ocean. It requires initiative to get off one's backside to find their obscure riches.

From that day on, my appetite was whetted to search out and discover the treasures hidden in other secret places. I was primed to believe in their existence. The world is not a drab place. As a pirate writes teasingly in the margin of his map, "there be treasure there," so it could be said of the natural and spiritual realm—there be treasure there. Such riches are only found by those who seek them and persevere. If I had never made the effort to find the small trout in the lake, I never would have stumbled upon the huge trout in the saltwater bay. One thing is sure—one never catches fish if a line is never put in the water.

I realized then that the things I might seek could elude me in the first places of investigation but then suddenly appear in another place, least expected, in a form much improved and quite different from my first conception. It takes the heart of a pioneer and an adventurer to press on until the gold is finally revealed. I determined to be one of those. I wanted to strike it rich. It wasn't only the fish that got hooked that day: I was hooked for a life of prospecting for the riches I knew were out there waiting to be discovered. This lesson would be repeated in one department of life after another, in both the natural and spiritual realms. I learned that I had only to cast off my bow lines to find myself in a new undertaking, with new tests and greater riches to be discovered.

Cast Off Your Bow Lines

Even in the later stages of my life, after many adventures and many treasures were sought and secured, just when I might have thought the time for such endeavours had ceased, a whole new chapter opened up and the call to venture forth came again. It all started with an affaire.

Cast Upon the Waters

PPSS
(of Poets, Philosophers, Sages and Saints)

Ship your grain across the sea;
after many days you may receive a return.
Invest in seven ventures, yes, in eight;
you do not know what disaster may come upon the land …
Whoever watches the wind will not plant;
whoever looks at the clouds will not reap.
As you do not know the path of the wind,
or how the body is formed in a mother's womb,
so you cannot understand the work of God,
the Maker of all things.

*"Sow your seed in the morning,
and at evening do not let your hands be idle.
for you do not know which will succeed …
or whether both will do equally well.*

—Ecclesiastes 11:1–2, 4–6

Still catching trout in the same place—fifty years later

7

2. An Affair

It was the beginning of the affair. There is no doubt looking back. It took a while to become a full-blown, life-altering passion, but the hook was set when I first saw her. It was truly love at first sight. With that first view from a distance, her beauty evoked a longing for something fresh to re-awaken my life. It felt like a call of the wild from a mystical place deep down in that primal somewhere most men carry within—a recklessness that wanted to throw off responsibility and cast myself into an exotic adventure. A part of me wanted the intoxicating rush of a fling—you know, "Damn the torpedoes and full speed ahead!" It must have been brooding in my subconscious for some time, awaiting the catalyst to turn it loose. I recognized this elemental unction that had taken hold of me at other times along life's journey and that totally altered the natural course of my life.

An affair is usually a destructive and essentially selfish chapter in a person's life that involves abandoning faithfulness to long-term relationships. It's invariably about yielding to the volatile passions lying just under the surface of an otherwise placid exterior. Often, it's a way to escape boredom and the seeming mediocrity of living the straight life—the patient fulfillment of everyday responsibilities to people who depend on your support and love. An affair can be a rebellion against the non-gratifying grind of life's drudgeries and the dullness of ordinary people—including, of course, oneself. Not so nicely put, it can be about just giving in to our lusts.

On the other hand, perhaps there can be an affair within the bounds of propriety, a legitimate expression of our deep desire to live life to the full and push beyond secure boundaries into the realm of adventure. It can be a grand passion—identifying and directing the deep inner motivations that make us feel alive. My scrabble dictionary has two definitions of affair. The one spelled with an "e" (affaire) is defined as a brief amorous relationship. Note the word "brief." It's not a long-term commitment. It is bound to be repeated, often many times.

This spelling probably indicates a French origin of the word. The second "affair" is defined as "anything done or to be done."

"Affaire" presumes sexual indiscretion and carries with it the accompanying shame and notoriety. The other "affair" is more simply an act flowing out of the essence of what makes a person tick. This kind of affair can be very legitimate and exciting, especially if it lines up with the inner motivations or passions that make us human. People were created to be passionate.

This affair in which I was entangled was to seriously alter the course of my personal journey. It came about late in life, after fifty-five years of living, including thirty-five years of marriage, a career, and raising a family of four. At the time I had no idea that it would add a whole new chapter to a life I thought was pretty well defined and played out. Turns out that my chance encounter with the beauty mentioned above brought a change into my life that I could never have foreseen.

Understandably, my spouse was also deeply affected by this intrusion of a third party into the normal course of our lives. Her whole life was altered as well, as you might expect in the event of an affair. What you might not expect is that my wife was an integral part of the tryst. We were both equally allured. In fact, Lyza first sighted our lady of the sea, peacefully afloat at a dock in Shearwater, a tiny coastal community halfway up the British Columbia coast.

"Would you like to come aboard and have a look?" The man, to our embarrassment, suddenly popped out of the companionway and caught us ogling the lady's lovely lines from the close quarters of the dock. We were practically caressing her teak topsides.

"Uhh … no … sorry, we were just looking … just admiring her. What a beautiful boat!"

He regarded us with a bemused smile and just a touch of smugness, like a man who knows he has the most beautiful woman in town.

"She is indeed. Would you like to have a closer look? Come aboard if you like. You can have a look inside."

"Really? We don't want to trouble you. Actually … yes, we'd love to see your boat."

We clambered aboard and descended a curving staircase of solid teak into the pilot house, which was also the galley and dining area. We gazed around at the spacious interior and then wandered through living quarters filled with hand-crafted teak, admiring the solid timbers and careful joinery of this marvellous exotic wood.

"Do you folks live on this boat?"

Cast Off Your Bow Lines

"We do now. We have a warehouse near Victoria where we keep a lot of our stuff, but for the last few years, *Porpoise* has been home."

Surveying the scene before me, I wondered how a chap like me could ever afford a ship like this. As though divining my thoughts, the man, whose name was Jim, volunteered:

"We charter her."

"What do you mean? Is she not yours?"

"No, we're the owners, but we do skippered charters. We take people out on her. They pay us to take them cruising on the coast. We feed them, take them fishing, explore the beaches, and teach them how to sail. It's great fun."

Right then a new thought planted itself in my mind. That seed would lay dormant for some time, but the germination began that day—there could be different ways to fulfill my passion for cruising the coast and impacting people's lives.

We continued our tour through the salon, staterooms, and the forward bunks. So much room—so livable. *Porpoise* was a lovely ship, a beauty of handsome construction and exquisite craftsmanship whose charms were as lovely on the inside as they were on the outside. She was a cutter-rigged ketch with two masts, four sails, and a spacious deck. For us country bumpkins of the yachting world, who spent our years on an old family boat of such rough and ready construction that its name was *Fred*, *Porpoise*'s charms were overwhelming. She was not only gorgeous—she was undeniably sexy. There was no doubt about it. If you beheld her from abaft, with her heart-shaped transom and large, jaunty Canadian flag aflutter, you simply had to admit that hers was the sexiest derriere in the harbour. Later, when she became ours, her attributes were constantly attested to by the regular parade of admirers passing by in their tenders.

"Nice boat!"

"I like your beautiful ship!"

"Is she a Bill Garden design?"

One young German sailor on his funky little sloop stood up with his arms spread wide in adoration as he motored by: "Vaht more could you possibly vant in life?"

We'd try to be nonchalant, but we couldn't hide our pride of ownership nor could we blame people for their open admiration. No wonder we both fell so helplessly in love. No wonder we felt no compunction for having run off with her. She was the girl of our dreams, the perfect match, a late-in-life love affair. At least that's how I look at it. Others observing our fling might have suspected it was an affaire of the French variety, complete with reckless abandonment of

An Affair

loved ones and responsibility. They always knew we liked boats, but to run off with a fifty-one-foot ketch and start a charter business, committing ourselves to four months away from home every summer, was taking this hobby a bit too far. When we later sailed her to Hawaii and then to Alaska and back, they knew we had slipped beyond their horizons and taken things way too far: "What about the previous thirty-five years of steady, hard-working, land living? What about your responsibilities to your spiritual family? You can't just up and leave us."

We could understand their wonderments and questions. We were often in wonderment ourselves. Yet although it may sound trite, and though some may think otherwise, to us this affair was meant to happen. It was destined to be an integral part of our unfolding life story. In the past, other passions were at work in us, resulting in different kinds of adventures. We had no regrets—not about our precipitous move across the border into Canada, or our pursuit of an unorthodox launch into youth work. We doubted we would have regrets about this new leap of faith. Those who truly knew and loved us understood the call of this emergent life. What could we say ... we were just that way. Today, seventeen years after the beginning of this affair, we still feel the same romantic thrill every time we step aboard our lovely lady. It appears to be a lasting relationship. We made the plunge and we are still not sorry for it.

Cast Off Your Bow Lines

PPSS

"... Alas for maiden, alas for Judge,
For rich repiner and household drudge!
God pity them both! and pity us all,
Who vainly the dreams of youth recall;
For of all sad words of tongue or pen,
The saddest are these: 'It might have been!'"

Excerpt from "Maud Muller"
By John Greenleaf Whittier

Porpoise, our beautiful lady

3. An Unlikely Prophet

The familiar cold blue waters of Haro Strait rippled in the chop of tidal eddies under our bow. The October air signaled both the change of seasons and the imminent change in our own lives. Decisions had been made, plans were in motion, and Lyza, our elder son, Dave, and I were all keen with anticipation as we stood on the deck. Now it was down to some dealing. *Porpoise's* full keel sliced across the irrepressible flow of Haro's strong current, powered only by the three air foils of the genoa, main, and mizzen sails. The same dynamics of wind over wings that keep a plane in the air were making this snow-white dove of the sea fly over the waves toward any destination we might choose. In this case, we were headed for Bedwell Harbour in the Canadian Gulf Islands to discuss with the owners of this delightful bird the details of her purchase.

My hands gripped the wheel possessively as she slipped into the zone—that place of balanced speed and smoothness where the sails are trimmed just right, and the wind is on the beam. The ship then settles into a rhythm of harmony with the waves, surging from trough to crest, coursing along at hull speed—the maximum a boat can do according to its design. I felt at one with this ship, as though melded to the wheel and fastened to the deck. We were like a skillful horse and rider finding their stride and galloping across meadows of water. Never before had I felt so exhilarated under sail. The call of the sea sang in my ears and ran at a fevered pace through my veins. In my heart I knew this lady of the sea must be mine.

I wouldn't have even imagined such a remote possibility when Lyza and I first encountered *Porpoise* four years earlier. At that time, we were cruising to the border of Alaska on *Fred Free*, our twenty-eight-foot sailboat. My father, Fred, had commissioned the building of the boat in the sturdy style of a Norwegian double-ended pilot boat and named it after a character in the *Little Orphan Annie* comic strip. It was finished in 1939, just in time for his honeymoon—a perfect wedding present for my mom.

Cast Off Your Bow Lines

Fred saw a lot of living in his sixty-four years—including hosting Lyza and me on our honeymoon, and some of our children on theirs. Our beloved boat was rough and ready and familiar enough to give us the security to tackle the Inside Passage, but he was short on creature comforts—very short—and a far cry from the classy *Porpoise*, who was almost twice *Fred's* size. How could I have known then that the comely ketch moored to that float in Shearwater would end up being ours? At that time, neither the circumstances of our lives nor our finances fit into a scenario of romantic sea adventures aboard a graceful yacht. How did we get to this juncture?

The answer to that question is best understood in the context of what was happening in our vocational life. In the spring of 2000, Lyza and I took a year's sabbatical after pastoring in one church for twenty-eight years. During our time away, we met the Blohms and their good ship *Porpoise*. After various other adventures on our hiatus—including a month-long driving tour of the Maritimes, a couple of autumn months holed up writing a book in a woodsy cottage in Maine, and a three-month stint at a friend's condo in Maui—we came back to White Rock refreshed and excited about the next chapter in our church community's life. Unfortunately, through a series of difficult situations over the next couple of years, we found ourselves struggling to make sense of a new set of realities in our church life. Our spiritual family had always been a secure place characterized by love and unity, but it didn't feel like that anymore. Instead, we fielded misunderstandings, discontent, disappointing revelations, and betrayals. It was an unhappy time for many, but Lyza seemed particularly bruised by the breakdown of trust in even some longstanding relationships.

Daydreams of our brief encounter with *Porpoise* could easily find their way into her imagination and stir fancies of sailing into a fresh chapter of life. For a while Lyza would shake off these visions as escapism and wishful thinking. But they kept coming back. Finally she'd had enough and decided that she needed some clarity. Lyza tends to cut to the chase and talks pretty straight most of the time, even to God. That's just the way she is. He doesn't seem to mind. So one morning she prayed and put her cards on the table:

Lord, I can't seem to get this dream of life on that boat out of my thoughts. It's true I'd like to escape because I don't like what's happening in our lives right now. I feel like I didn't sign up for this. But I want to be in the centre of your will because I know that's the only place real happiness can be found. So if these thoughts aren't in line with your

An Unlikely Prophet

plans for us, please make that clear to me. On the other hand, if the reason these thoughts keep persisting is somehow tied up with your will for our lives, please make that clear. Could you give me an unmistakable sign either way so I know how to set myself going forward? And I am asking that you show me that sign today. Amen.

Later that day, Lyza came to me holding her day-timer and said she had something important to tell me. She told me about her prayer. I was about to suggest that it might be better to not put a timeframe on God, but—looking at her face—I thought better of it. Her planner was formatted to show a week at a glance with a quotation for each week at the top of the page. She read to me that week's saying:

> *"Twenty years from now you*
> *will be more disappointed*
> *by the things you didn't do*
> *than by the ones you did do.*
> *So throw off the bow lines.*
> *Sail away from the safe harbor.*
> *Catch the trade winds in your sails.*
> *Explore. Dream. Discover."*

She believed this was the answer to her prayer. It was the permission she needed to dream without guilt, to know it wasn't just escapism but a call to move forward, to pass through whatever doors might open, and to believe for more adventures ahead in our lives. The nautical terminology sealed the deal for her. It hit her as a direct word of hope from God.

I've learned through many years of living with my wife that she has a certain spiritual intuition that I sometimes lack. Seeing how glowing she was about it, I didn't want to dampen her enthusiasm, but I wondered if this dream lacked a little in the realism department. I tread carefully when she declared that she believed we would own *Porpoise* someday.

"You think we're going to have *Porpoise*? Don't you mean that maybe one day we'll have a boat like *Porpoise*?"

"No, I believe that somehow we are going to have that boat."

15

Cast Off Your Bow Lines

"But the Blohms love their life on *Porpoise*. She's their baby, their pride and joy. There's no way they would want to give her up. Besides, how could we ever afford a boat like that?"

Lyza looked at me with an air of certainty that only the true believer can manifest. She just shrugged and said, "I don't know how, and *I* certainly don't have any intention of trying to make it happen. All I'm saying is that I believe God is going to make it happen. When and how are up to Him."

To her credit, she never brought it up to me again. She just tucked it away in her heart and waited. I'm ashamed to say that this incident, and her special quotation, quickly evaporated out of my mind—a fairly regular occurrence, according to my wife.

About a year and a half later (plenty of time for anybody to forget, right?) I was sitting in my favourite chair and gazing out the window. It was definitely a Rodell Grey kind of day. A general contractor friend of ours had invented this custom tint for one of his lottery homes, and the moment our daughter, Gigi, saw it, she told us it was the perfect colour for our new seaside home. When the sun, wind, and clouds combined in just the right mix to make the light shine down on a turbulent sea, it produced this misty greyish-green hue that exactly matched the fresh paint on our living room walls. The day matched the mood of my reflections: sombre yet lightened by hopeful thoughts. The previous three years had been tough, but new prospects made our life feel a bit brighter. In fact, I hadn't felt so lighthearted in a long time. It probably helped that we were living, again, within close proximity to the water. Gazing on the sea every day for the past year had been rejuvenating for both Lyza and me—our mutual happy place. One row of houses up from the shore on the steep White Rock hillside, our perch was a daily reminder of how much the waters of coastal British Columbia were a place of nurture for our family. Here we felt separated from the stress and pressures that had drained and wounded us. Mesmerized by the sight of the endless wave trains rolling over tide pools on the flat sandy beach, the grip of the recent negative events in our land-life began to loosen their hold. I could hear again the siren call of the sea.

Maybe that's what got me thinking again of that chance encounter a few years back on a dock halfway to Alaska. I started thumbing through the rolodex. I remembered Lyza had oddly filed the card under P for *Porpoise*, rather than under the owners' names, but I found their information set down below. On a whim, I decided to call Jim Blohm. Once I had him on the line, I felt a little embarrassed calling him out of the blue. But I forged ahead.

An Unlikely Prophet

"Hi, Jim, my name is John Clarke from White Rock. My wife and I met you and Kathleen a few years ago on the dock at Shearwater. You probably don't remember us. We were the ones with the adopted daughter from Haiti," I said. People never forget Ginny. "We came on board *Porpoise*, and you showed us around."

"Oh, yes, sure. I remember you." Jim seemed very friendly and open to talk, so I got right down to the reason for my call.

"Well, this might seem a little strange, but I was wondering if you'd be willing to have lunch with us sometime. We want to know more about your chartering business and how it works on a boat like *Porpoise*. One day we might be interested in doing something similar."

"Really? Sure, we'd be glad to meet with you."

We set a date for the following week to meet at a little café for lunch on Vancouver Island. As I was wrapping up the conversation, he threw in, "Maybe I should mention—it looks like *Porpoise* may be for sale at the end of this summer."

This was not what I expected to hear. Not at all. Somewhat flustered, I blurted out the first thing that came to mind. "How much?"

"A hundred and seventy thousand US," he answered right back.

This wasn't a very smooth interaction, but I was so shocked to hear she was for sale that blurting out the question of price and hearing his response put me in a kind of stupor. The first thought that rolled around in the back of my head as Jim kept talking about *Porpoise* was, *That's not so bad. Heck, I could buy a small condo for that price, and this floats ... and sails.* I'm not sure what I had expected. I had no idea of the comparative prices of boats on the market. Jim could have said half a million and I wouldn't have blinked. *Porpoise* seemed like a whole lot of awesome, teak-covered floating wonderfulness for that kind of price. And coming up with that amount of money was actually doable. I didn't have it just sitting around, but our house was paid for and we could take out a mortgage on it. In a moment, the idea of owning our dream boat had become a possibility.

Lyza had been out shopping when I set up that get-together, and I had never indicated to her any interest in contacting the Blohms. She might have had something to say about the matter, since it looked like *Porpoise* was soon to be available. Admirably, when I told her I had made that call, there wasn't a whisper of "I told you so." She just kept her own counsel and didn't throw in her two bits, as she usually would. I was so engrossed in my own thoughts that I was just glad she didn't question my unilateral action in setting up the meeting.

We met Jim and Kathleen Blohm for lunch as planned, and since we had the sailing life in common, we hit it off right away—as cruisers on the coast

Cast Off Your Bow Lines

often do. They had lived aboard and chartered for six summers and had a slew of interesting stories to tell. The lunch hour passed quickly, and they answered our many questions with both patience and enthusiasm. They clearly loved what they were doing, which begged the question: Why were they considering selling their beloved boat? Their answer was that circumstances were changing for them and they were building a house on some property up the coast at a place called Sointula. They had other dreams to follow there, including a studio for Kathleen, who was an artist. We filled them in a little about being in transition ourselves and unsure about our next steps.

As Jim walked us to our car at the end of our time together, he said, "Well, we all have a lot to think about, don't we? We're planning on taking the boat up to Sointula this summer to work on our house. It's hard for us both, especially Kathleen, to let go of the boat, even though it allows us to fulfill other dreams. We're in no rush. Why don't we get back together in the fall? Maybe by then we'll all be more settled about what's happening in our lives."

We were relieved to have some time to process it all. We needed that time for the many conversations that had to take place between us and the other people in our lives before we could make any future plans about *Porpoise*.

So here I was, several months later, at the helm, with all the discussions and decisions done. Lyza, our son, Dave, and I were delighted to be aboard as we crossed Haro Strait on this crisp October day. To say I was smitten would be a gross understatement—I was totally infatuated. The irresistible pull of the flood tide carrying us along toward South Pender Island enhanced the feeling of being irrevocably drawn into a fresh new chapter. There was more in the wind than the cool evergreen air of the Pacific Northwest; a sea change was happening.

The lovely ship rounded Wallace Point. We dropped the sails and anchored off Poet's Cove in Bedwell Harbour. At that point, Jim and Kathleen had owned *Porpoise* for about eight years, having picked her up in a bank sale when the former owners had become insolvent. The last owner had let her fall into disrepair, and the Blohms lovingly updated her systems, refinished her teak brightwork, and restored her to her former glory. Having put in so much sweat equity to get her into pristine shape, it was hard for them to let her go. But when we set up this meeting on the phone, Jim explained that they hadn't had time to take *Porpoise* out even once that whole summer. That's what had finally convinced Kathleen it was time to sell. She'd said to Jim: "She's like a thoroughbred. She has to run, not be left in the barn." To us, it felt like they had done all this work to get her ready for our new life.

An Unlikely Prophet

Kathleen bustled about in the galley for a few minutes and then set before us plates and coffee cups. As the rest of us found spots around the table, Lyza picked up the cup in front of her and commented on it. "I love the photo of *Porpoise* on the cup. Did you have these made, Kathleen?"

I picked up my cup to examine it, but instead of a picture of the boat, mine had a quote on it. I read it aloud:

> *"Twenty years from now you*
> *will be more disappointed*
> *by the things you didn't do*
> *than by the ones you did do.*
> *So throw off the bow lines.*
> *Sail away from the safe harbor.*
> *Catch the trade winds in your sails.*
> *Explore. Dream. Discover."*

As I was reading, I had the vague thought, *This sounds familiar. I've heard this somewhere before.* After I'd read the quote aloud to the end, Dave exclaimed excitedly, "That's it! That's the quote … isn't it, Mom?" I still hadn't clued in to what he was talking about, but one look at Lyza's face brought it all back in a rush. Tears were brimming in her eyes.

I felt that prickly sensation you get when something spooky happens. I remembered. The day she'd read that quote was the day she also confidently declared her belief that eventually we would own *Porpoise*. Now here we were meeting to finalize the details on her purchase. I'm pretty sure I'd never heard the saying before, except for these two related instances, and the "coincidence" seemed serendipitous. The first time the quote came up was for Lyza, but this one was for me, I think, to assure me that we were on the right track.

By this time, any businesslike atmosphere was totally shot as our unguarded exclamations and emotions spilled all over the table. Our cards were certainly not being played close to our chest—the way you're supposed to act when trying to make a deal—so there was nothing to do but tell the whole tale. We were already aware that the Blohms adhered to a more Eastern spiritual philosophy, so I wasn't sure how they would respond to Lyza having prayed for a sign. At the end of the story, Kathleen replied, "Well, this sounds like it was meant to be." I felt gratified at her gracious response. Of course, I also think she was gratified

Cast Off Your Bow Lines

to know her precious *Porpoise* was going to a loving home where she would be properly appreciated.

By the time we hauled anchor and left Bedwell, a deal was struck, and we were to take possession the following May, giving us several more months to get all our affairs in order. The icing on the cake occurred as we motored the boat back across the strait. Again, I was at the helm, revelling in the knowledge that I was to be captain of this lovely vessel. The sun was setting over Portland Island in a blaze of orange and yellow against the darkening blue of the evening sky when a familiar blast of air escaped strong lungs and a slapping splash sounded first in my right ear and then in my left. I knew what it was immediately. "Porpoises!" I shouted.

Dave and Lyza rushed up to the bowsprit, knowing that's where these spirited mammals would head. This pod of a dozen began their rollicking antics under the bow, and Jim graciously relieved me at the helm so I could go watch them too. The group of Dall's porpoises graced us with a real circus act, complete with leaps and breaches and torpedo runs across the bowsprit. As I leaned over to get the best view, Lyza came up softly behind me, put her arms around me, and whispered in my ear, "It's another sign, you know. We're meant to have *Porpoise*."

Seven months later, our entire family piled on board our awesome new acquisition for the voyage to her new home at a marina near White Rock. It would be hard to find a more enthusiastic bunch as we roved on deck and below to discover all *Porpoise's* many attributes. We were in possession of a fifty-eight-foot sailing ketch, a Bill Garden design, Hudson Force 50—one of the well-known marine architect's favourite designs. We were her fifth owners and proud to have her. She was thirty-five years old, having been built in Taiwan in 1975. She'd crossed the Pacific Ocean a couple of times, and now she would be in coastal service to our family and those who wanted to cruise the British Columbia coast with us.

Since making our deal with the Blohms the previous fall, Lyza and I had upgraded our credentials, both of us becoming certified coastal skippers. Later I was to add sailing instructor to my qualifications in order for *Porpoise* to become a recreational sail training vessel. Our plan was for her to be a working girl and pay for her own upkeep through charters. But this was also my greatest vulnerability in fulfilling our dream. Would people really pay to go sailing with us? Were we really up to this chartering business, where we would have to skipper, serve up great food, clean up, be host and hostess, teach people to sail, *and* be recreational directors? This could prove to be a daunting new vocational enterprise for two people already in their fifties.

An Unlikely Prophet

Nevertheless, we were full of enthusiasm for a fresh chapter in our lives. We arranged to continue on church staff during eight months of the year in a support capacity as new church leadership rose up. During four months of spring and summer, we would (hopefully) support ourselves with charters. And as an added way to make our salary, we would live on the boat from May through September and rent out our house as a vacation rental. It all seemed sound in principle, but we still needed to see if it would pan out.

During that first summer of ownership, June quickly passed into July, but nothing materialized in the charter department. As part of the deal, Jim and Kathleen had passed on to us their charter website, *Pacific Encounters*, but there wasn't much action from that quarter. Since we had no charters booked, we simply made *Porpoise* available to our friends and church family for afternoon and overnight cruises. Lyza kept a record that first summer, and we ended up having at least 184 guests on board by the end of July. Of course, there was no money in that, but we all had a great time. By August, though, I was really starting to get worried. I wondered if I had just acquired a twenty-six-ton millstone around my neck. To be able to afford the upkeep, moorage fees, and all manner of expenses that come with owning such a boat, we needed her to earn her keep.

Lyza, as usual, was much more upbeat about the situation. She remonstrated with me one day when she heard me fretting: "Don't you think if God got us into this, He's going to provide what we need to make it work?"

"I know that's how it's supposed to work, but we're already halfway through the summer and haven't even had a bite. People plan trips like this way in advance. Who's going to do a big sailing trip in August on the spur of the moment?" I felt bad about my unbelief, but I was not by nature a big risk taker financially. Like Cuba Gooding Junior in the movie *Jerry Maguire*, I wanted to say, "Yeah, but show me the money!"

The last week of July found us anchored with friends in Vancouver's English Bay to watch the summer fireworks display. Most of the myriad of boats took off after the finale, but we stayed overnight to avoid the crowds. The following morning, we were all at the breakfast table when Lyza's cell phone rang. When she saw that it was a call from Orca Boat Charters, she went topside to take it. I went too, anxious to hear what it was about.

"Hello, my name's Michael. I'm wondering if you folks have any space for a charter in mid-August somewhere in Desolation Sound."

Now the truth is, we were so green at this stuff, we didn't even get that this guy was a broker—a kind of a middleman between a charter yacht and a

21

Cast Off Your Bow Lines

potential customer. No matter, Lyza is quick on her feet in situations like this, and she realized that whoever this guy was, he was offering us our first charter. She played it pretty cool, I thought.

"Let me check my calendar," she said, shooting me a significant look and raising her eyebrows. "So what are the specific dates you have in mind?"

When he gave her the dates, she calmly inquired, "So you're wanting a ten-day charter?" Another arch of the eyebrows my way. After listening to his reply, she said, "Yes, I think we can manage those dates. We'd need to reposition ourselves up to that area, but we have time to do that."

I was marvelling at Lyza's businesslike manner. *Where's my wife and what have you done with her?* I thought. This was another of those moments that reminded me never to underestimate this woman. I moved in closer to the phone, hovering like a vulture, to hear what Michael was saying back to her.

"The family is from back east, a father and five children. We're also looking at some other options as well, including *Copper Sky* in Alaska. If you're interested, a representative of the family's foundation would come to interview you to make sure it would be a suitable arrangement for them. I've sent the pertinent information to your e-mail address as well. I'll be in touch to set up an appointment."

When Lyza finally hung up the phone, we just grinned at each other. I knew such an extensive charter would go a long way toward the bottom line and be a bigtime start in the charter business. But hearing that "looking at other options" comment, I was afraid to count my chickens before they hatched—especially after we got back to White Rock, and Lyza looked up the "other option" Michael had mentioned. Even Lyza looked a little abashed when she saw that *Copper Sky* was an eighty-eight-foot schooner in Alaska with staterooms to spare. *Porpoise's* accommodations couldn't compare, and we knew it ... until we read the e-mail Michael sent us. It contained all the info he had mentioned on the phone, but at the bottom of the page was this:

> *"Twenty years from now you*
> *will be more disappointed*
> *by the things that you didn't do*
> *than by the ones you did do.*
> *So throw off the bow lines.*
> *Sail away from the safe harbor.*
> *Catch the trade winds in your sails.*
> *Explore. Dream. Discover."*

An Unlikely Prophet

Lyza literally gasped but then quickly got over the shock. She pointed to the quote and said, "Well, I guess we have our first charter no matter what the odds, don't we?" And that was the end of any doubt for her. I have to admit that after feeling my hope somewhat dashed by the sight of *Copper Sky*, the special quote showing up so unexpectedly had nailed to the wall any of my remaining unbelief. Doubting Thomas lingered in the back of my head until I saw some cash in hand, but having received that quote for the third time, I was ashamed to admit he'd been there at all.

Michael duly sent a representative from the family foundation to interview us. As the rep explained the family's situation in more detail, we understood why this charter had been booked last minute and also why it was being handled with such thoroughness. The family had recently lost the children's mother, who had suddenly and tragically died only a couple of weeks before. They were looking to get away and spend some healing time together in our coastal environment. The man they sent was obviously compassionate and solicitous of the family's welfare and actually seemed pleased that our background included pastoring. His comment at the end of our conversation was telling: "I know there may be more luxurious charter possibilities for this family, but given these difficult circumstances, I think you're just the right people and this is just the right boat for this situation."

Within a few days, we had a 50 per cent deposit on the trip and were launched into the charter business. The following days became a riot of frantic preparation, but we managed to put the best foot forward in our fledgling business. We actually had a fair bit of confidence for this charter, considering our background With our fifty years of boating, we knew these cruising grounds like the back of our hands, and with our pastoral experience, we were old hands at hospitality and had the skills to deal sensitively with people who were grieving. It turned out to be a very significant and affirming trip for all of us. It was the first of many "divine appointment" charters that would materialize in the ensuing years.

Cast Off Your Bow Lines

PPSS

> Serene, I fold my hands and wait,
> Nor care for wind, nor tide, nor sea;
> I rave no more 'gainst time or fate,
> For lo! my own shall come to me.
> I stay my haste, I make delays,
> For what avails this eager pace?
> I stand amid the eternal ways,
> And what is mine shall know my face.
>
> Excerpt from "Waiting"
by John Burroughs

Mark Twain on the deck of a ship

Book Two

Sail Away from Your Safe Harbour

A Note to the Reader

Pacific Encounters was the name of the charter business we inherited from *Porpoise*'s previous owners. It was a good name. What better descriptor for our endeavour than an up close and personal encounter with this Pacific coastal realm we had cherished through many years of family boating. The marvel to me in acquiring *Porpoise* was that what had always been such a private treasured experience could now be shared with others. We could squire other people around our secret world of harbours, coves, inlets, and beaches—all the way from the Gulf Islands of British Columbia to the border of Alaska. I had to admit that a part of me was almost reluctant to share it with others. Being able to live the vagrant gypsy life for a few precious weeks a year was always our privileged escape from our harried land life. There's value in keeping private certain places in your soul and world.

On the other hand, keeping your treasures hoarded is no fun at all. Life is meant to be celebrated. Letting our stories and experiences mix with the stories and experiences of our fellow sojourners on earth has got to be one of the purposes for which we are here. With *Porpoise*, we had a ship capable of taking five to six other people along with us in relative creature comfort. With our newly created website posted on the internet, literally anyone on the face of the earth could find us and book a trip. Not only was this destined to be a *sailing* adventure but a *people* adventure as well. We wondered whom we would meet. We looked forward to discovering their life stories, and we anticipated making new friends and broadening our contacts with others from different parts of the world.

That is exactly what happened right from the start. We made connections with people we wouldn't have otherwise known even existed—from as far away as France, Sweden, New Zealand, and England—as well as from all over the United States and Canada. The following narratives describe a couple of those skippered charters during our first five years on *Porpoise* and also some family excursions (including one involving our beloved cat, Mr. Miyagi).

One more word of explanation: Because British Columbia summers in the lovely, protected waters behind Vancouver Island are generally warm and pleasurable, the vast majority of our family trips and charters have also been warm and pleasurable, a relatively tame walk in the marine park. The wilder adventures are more likely to be found travelling farther afield to the open ocean west side of Vancouver Island and to the central and north coasts of BC. The more harrowing experiences are clearly more interesting to read about if not to live through, so they are mostly the ones included in these chronicles.

The even greater harrowing experience was to leave the safety of the coastal environment to sail the great Pacific Ocean. This became our own raising-of-the-bar Pacific encounter when we sailed to the Hawaiian Islands in 2010. The journals we made over those three months comprise the third section of this book. But first a sampling of coastal adventures …

Pacific Encounters brochure

For more information about Pacific Encounters Sailing Adventures,
see the website

www.pacificencounters.com

4. Bernie

My new friend Bernie gazed up into the log and plank ceiling sixteen feet above him and then scanned about the open interior of our Nelson Island cabin. Nothing but wood covered walls, floor, and roof. Even the furniture was made of wood. Pine, maple, birch, hemlock, and, of course, fir and cedar were all represented, the natural provision of the twenty-five-acre forest surrounding us. It wasn't your average wood frame structure made with the usual framing techniques. It was built of logs cut from the property and assembled post and beam style with two-by-six framing in the spaces between the upright logs. It certainly wasn't a perfectionist carpenter's dream ... maybe his nightmare, if you looked too closely at the joinery ... but to most eyes it was an amazing habitation that seemed to have grown out of the forest. It even boasted a curvy arbutus tree stairway (with branches and all) that created a whimsical feel in counterpoint to the tall, straight fir trunks beside it supporting the roof.

The cabin was cozy and functional, a place to hide out far away from the crazy intrusiveness of our modern world. Everything we needed for basic living was available. We had good water pressure in the taps because of the gravity flow in the pipes down from our water source, a spring up the hillside. We had hot showers and a sauna. The little bathroom was a small separate structure on the hill behind us we called the "looky loo," because sitting upon its flush toilet we could look out over the inlet through the trees. Lyza and I had worked together all one summer vacation building the stone hearth made entirely of hand-picked boulders and rocks from our own beach just out front. The wood stove set into that hearth kept the cabin comfy warm in cooler weather. I never tired of arriving and first stepping into the cabin, breathing in the redolent scent of cedar that filled the wood box and covered the walls. Bernie surveyed it all with evident satisfaction before plunking himself down on the forest-green futon and regarding me with a bemused grin.

"So you built all of this yourself?"

Bernie

I could tell he was sizing me up as he had been for the previous three days of our chartered cruise. The whole coastal environment was a new and wondrous experience for him and his family, and now this sampling of wilderness living in a rustic cabin, isolated as it was in the midst of acres and acres of mountainous crown land, introduced another layer of contemplation. The only access to the cabin was by boat, and *Porpoise* was anchored and stern-tied in the little cove out front, a tiny indentation in the miles of rocky coastline surrounding us.

"Yeah, I built it with help from my family and a little help from my friends. It took us ten years of vacation time to carve out our little getaway here. We wanted to use as much building material as we could right from the surrounding land. We cut down trees, peeled and limbed them, milled up timbers with a mini-mill, and hauled them into place with block and tackle. The only power tool used was a chainsaw. Then we scrounged up used stuff, like windows and doors, to keep down the cost."

Bernie shook his head, and said, "That must have been a lot of work!"

"Yes and no. We took our time. Since it was all done on our summer holidays, I'd work in the mornings on some project and then play with the kids in the afternoons, either swimming in the cove or fishing. It was probably the most fun recreational thing I've ever done, with the combination of getting to enjoy this beautiful wilderness and accomplishing something lasting for my family at the same time."

"I could see how that could be very satisfying, especially getting to work with your children. Everyone would be truly invested in the project."

I nodded enthusiastically. "It was very therapeutic to build something physical here and know that it was nailed down and would still be standing when we returned. I didn't often feel that way about my work back home. When you work with people, it's easy to question whether what you're doing is having any lasting effect."

Bernie looked at me quizzically. I think he had been waiting for a natural opening to ask a question.

"So just what did you do before—or, should I say, when you're not doing this boat captain thing? This sailing business has to be a seasonal thing in these waters. What's your line of work most of the time?"

I kind of expected this question would come up sometime in the trip. Generally, Lyza and I tried to travel "incognito" (relative to our former life). We figured people wouldn't jump at the chance to take a vacation with a pastor. In other circumstances we had found that when people discover you're a minister, they

29

Cast Off Your Bow Lines

can become ill-at-ease and start acting differently. I call it the "halo effect." Since one of our goals on coastal cruises was for people to let their hair down and relax, we didn't want them trying to clean up their language or becoming self-conscious because of us. Much better to be perceived as the normal types we really are. Later, after people got to know us, our background wasn't an issue. The seemingly incongruous blending of pastor and salty sea captain became just another point of interest to add to the list of first-time experiences they were having.

It was the halfway point in this particular cruise, and Bernie was beading in on me. I met his eyes with a twinkle in mine.

"No, this wasn't what I did for my life vocation. What do you think I did? Try to guess."

Bernie hesitated at first and then decided to play the game.

"Hmmm ... let's see. Were you some kind of businessman?"

"No. Well, I guess I was ... sort of." I was thinking of the various business skills required for running a big church. For years I had set and overseen budgets, run annual general business meetings ... that sort of thing.

"How about a doctor? Were you in some kind of medical field?"

"No." But then I thought of all the counselling one does as a pastor, trying to heal people's wounded hearts. I suppose I was a doctor of sorts. Healing was a part of my role. After a pause, I said, "Again, sort of, but not in the way you're meaning." I could tell Bernie was intrigued.

"Then how 'bout a teacher?"

I knew he was hoping, as I didn't answer immediately. As I thought of the many classes, seminars, and study groups I'd led in my career, not to mention the weekly sermons, I knew being a teacher was definitely part of my vocation. Luckily, he didn't wait for an answer but jumped in with, "Were you a lawyer?"

"My dad was a lawyer and wanted me to be one, but I didn't go that way. On the other hand, I was a lawyer of sorts. I handled the greatest law book of all."

"I don't know. I give up. What did you do?"

I knew it was time to 'fess up.

"Okay. Well ... for over thirty years, Lyza and I have been pastors."

Bernie lay back on the futon, staring at me with a blank expression.

"A what? Pastors? What do you mean?" He still wasn't registering. "You mean like a priest?"

"Yeah, like a priest. A guy who leads a church, preaches, does weddings and funerals ... that kind of stuff. We were part of a fairly large church in White Rock."

30

Bernie

At that, Bernie did a classic jaw drop, with eyes widened and eyebrows raised. This possibility hadn't even remotely crossed his mind. Somewhere in his French ancestry were Roman Catholic roots, but the celibate priesthood was obviously not a slot he could fit me into. He shook his head and said, "I never expected you to say that." I could see the wheels of his mind turning over the last three days as he checked over his language, among other things.

After a little more clarification, Bernie regained his composure and, over the next few days, seemed to actually relish this new revelation. A new topic of conversation had opened up, and he had a lot of questions. After all, he was imminently facing some eternal questions. The doctors had discovered that he had thoracic cancer and had given him only a year to live. When he first contacted us over the phone, he was quite forthcoming about his condition and prognosis. This sailing trip was one of the things on his bucket list. He said he was from Kelowna and that, although he had a boat and had done lake sailing, he had never sailed in British Columbia coastal waters. It was his dream to take his partner, Sherri, and his nephew's family on a sailing trip.

When he showed up that summer, we would have never guessed he was sick, let alone suffering from a terminal illness. Tall and lanky, he appeared energetic enough as he clambered aboard to give us a ready grin and a firm handshake. Behind him came Sherri and the rest of his crew of relatives—a nephew and his wife with their teenage daughter and her boyfriend. Along with their luggage, Bernie hefted aboard a big cardboard box chock-full of wine. This was not just any wine but special vintages and brands that he'd acquired while biking through the countryside of France on a winery tour. We learned later that he had sent similar boxes to his brothers and sister, along with instructions on when to drink which bottle. That way they could celebrate life together even though they were in different geographical locations. Obviously, he wasn't the type to face his looming demise with a morose spirit. Bernie intended to live his last days on earth in style.

With the sense of bonhomie that radiated from him, it wasn't hard to feel an instant affection for this man. Looking beyond his cheerful countenance, it was plain there was also a well of sorrow. His eyes, a liquid brown, exuded both a pathos that revealed the tragedy that had befallen his life and a sparkle that revealed his love of life. He was curious about everything, and as soon as he learned of our roots, his questions about our life story and spiritual journey were penetrating and full of genuine interest. As time went on, he was as keen on spiritual discovery as he was on coastal sailing discovery.

31

Cast Off Your Bow Lines

Lyza and I were left to marvel over the apparent synchronicity of our past and present calling. Through chartering, we were discovering new people, sharing their lives—both the joys and the sorrows—in a very comfortable way, swapping our life stories, and learning from each other. Our passions for cruising and for helping people on their life journeys were fused together in an enjoyable mix that included both the wild wilderness of our coast and people's wild and woolly life stories.

We came to like the adventure of not knowing whom we might meet next. When guests came aboard, their ordinary lives were temporarily suspended. Cell phones were laid aside, and the passing of time slowed down to embrace the here and now. The only "business" to transact was food and fun and encounters with nature at its most pristine. With lots of time to talk and share unique experiences together, building new friendships was natural and gratifying to all, both skippers and crew. It soon became apparent to us that Pacific Encounters was the perfect name for our business.

The rest of that week with Bernie and his family passed all too quickly as we sailed them up to Princess Louisa and back to our cabin site on Nelson Island. No one wanted it to end as we travelled down Agamemnon Channel to Pender Harbour, but our time had run out. Before we knew it, we were about to say our goodbyes as we gathered in the cockpit just before running them ashore in the tender. I decided that I had to say something before they parted.

"We've so enjoyed spending this time with each one of you. Just before you go, we wondered if it would be all right to pray with you. I already asked Bernie if this would be okay, and he said it would be fine."

The teenagers looked a little surprised, shooting a glance at each other. It could have been awkward at that moment, but somehow it wasn't. Although the subject of cancer hadn't been brought up much, everyone was aware of Bernie's terminal prognosis lurking in the background. It was there as a dark backdrop on a stage where vivacious Bernie played the central role in a wonderful and memorable week. But now they were all heading back to real life—to all the pressures, concerns, and doctor appointments.

The family sat crowded together in a quiet semicircle on the boat's cockpit bench, wondering what I was planning to do. I was wondering the same when, to my surprise, I dropped down on the deck on one knee and laid a hand on Bernie's chest. I'm not much given to such overt demonstrations, but I was emboldened by the circumstances and the trust that had built up between us. I didn't care if it appeared a bit radical. I looked straight into Bernie's eyes.

Bernie

"Bernie, I want you and Sherri to know that whatever lies ahead with you, if Lyza and I can help in any way, we want to be there for you. We can come up to visit you or whatever you need. Just let us know."

Tears slowly ran from the corners of Bernie's eyes and down his cheeks. I glanced over at Sherri and saw tears on her face also. I thought to myself, *Nobody here asked me to be their pastor, but then again, I didn't need to be in a church building, and they didn't need to be church members for me to care about them or pray for them.* It was simply about loving the person beside us.

"Father," I began, "In Jesus' name, I ask you to bless Bernie and his family. I ask you to heal him body and soul, to arrest this cancer and preserve his life upon the earth so he can continue to be who he is to the people around him. Open the eyes of his heart to see you better in the world around him and to discern how you are at work in his life."

I glanced around. The tears continued to flow. Even the kids appeared to have survived the prayer. We all stood up and hugged each other, aware that this was one of those special experiences that would not be forgotten by any of us. I ferried them ashore in the tender, said goodbye, and they were gone.

After the season, Lyza and I drove up to Kelowna and spent a weekend with Bernie and Sherrie, furthering the special connection. The next spring, we got a call from Bernie.

"Let's go on another sailing trip this summer," he said in his usual enthusiastic manner.

"Hey, Bernie, how you doing?" I said. "I thought you were supposed to be dead by this summer."

"Well, I guess I'm not. We just returned from a winery tour bicycle trip through France."

"Doesn't sound like a person on their last legs to me! You must be feeling good. If you're up to it, we'd love to do another trip with you."

"I'm feeling fine. I'm still getting tired and have to take a rest in the afternoons, but I don't feel much different than this time last year."

He came out with us again for a September cruise to Desolation Sound. He came by himself this time, and we enjoyed his company and conversation during a streak of fine Indian summer weather. He talked about how his perceptions of the role of God in his life had definitely grown. He told us that he and Sherri had even started attending a thriving church in his community. Before the end of his September cruise with us, he was already talking about a third one the following summer. The one year that the doctors had given him now stretched into three,

33

Cast Off Your Bow Lines

and it looked like he wasn't slackening his pace. I dared to hope he had dodged the bullet, but the following spring I got a troubling phone call.

"Hi, John. I'm looking forward to our sailing trip this summer and wanted to fill you in on what's happening on my end. Right now I'm in the hospital for an experimental treatment with radioactive beads in my liver. I'm not feeling very good at the moment, but I'll call you when I get home and we'll work out the details for our upcoming trip. I could use some of your power prayers right now."

My heart sank as I hung up the phone. I had witnessed too many cancer patients suffer the effects of radiation or chemotherapy ... and now radioactive beads? I had come to wonder what killed people faster—the cancer or some of the treatments. Perhaps the jury's out on that question, but the dread I felt upon hearing this would not go away. We thought he'd been doing so well. Why impose some experimental treatment? Maybe too much money exposes one to too many options. Whatever the case, two weeks later I got a call from Sherri. Bernie was dead. Just like that. Gone. I struggled with anger about it but eventually let it go. All of us are on borrowed time. I still think of my friend, and Lyza and I talk about him often with great fondness. At least he went for it. He lived to the hilt. We were glad to be able to share that with him, to be a part of his pacific encounter.

Bernie

PPSS

"Tell me not, in mournful numbers,
Life is but an empty dream!
For the soul is dead that slumbers,
And things are not what they seem.
Life is real! Life is earnest!
And the grave is not its goal;
Dust thou art, to dust returnest,
Was not spoken of the soul …
… Let us, then, be up and doing,
With a heart for any fate;
Still achieving, still pursuing,
Learn to labor and to wait."

Excerpt from "Psalm of Life"
By Henry Wadsworth Longfellow

5. Miyagi in the Belly of the Whale

Mr. Miyagi balanced precariously on the edge of the interesting cavity that my friend Dwight and I had opened up in the floor of the pilothouse. We were doing some maintenance on the big red diesel that hunkered down in that dark cave. The cat had no idea it was the source of the horrendous roar and subsequent terror that plagued him in his new life aboard *Porpoise*. In its dormant state, it was just another of the incomprehensible objects to reconnoiter here. He stepped gingerly along the teak parquet floor, keeping perfect balance on the edge like a tightrope walker, and peered intently down into this whole new world of wires, pipes, metal, boxes, and hidey holes—all of which he had been unaware of until now.

"Don't even think about it, Meegs," I warned, growling in his direction as I fumbled with a drain nut underneath the governor. His impassive yellow eyes met mine as I tried to impress upon him how ill-advised it would be to embark upon an investigation of this particular space. The engine under his gaze seemed innocuous enough while at rest, but …

"I guarantee you would not like to be intimately acquainted with Mr. 120 horse Ford Lehman. Believe me!"

The tip of his tail twitched back and forth behind his inscrutable face, usually a sign of impish and unpredictable behaviour to come. The phone rang and our attentions were diverted to other matters. A few minutes later, we slid the heavy wood engine covers back into place. It was time to get going. Some friends were meeting up with us at the White Rock pier. Lyza and I were slated to take them out on the water for a weekend on *Porpoise*. It was always with pardonable pride that we sailed guests across the Salish Sea (aka the Strait of Georgia) in our magnificent ship to visit our favourite coves and beaches.

Later that afternoon, halfway across the strait, Lyza popped her head up the companionway while I was in the middle of a genial conversation with our friends and asked, "John, have you seen the cat?"

Miyagi in the Belly of the Whale

"Uhh … no … not for awhile. He was definitely on the boat this morning."

"Well, I'll keep looking, but I haven't seen him anywhere. You don't think he could have been left behind at the marina in Blaine, do you?"

"Naw, I doubt that," I said confidently, but a slight warning niggled in my mind. I didn't think he'd been left on shore, but he did have other tricks. While on board he could disappear from sight periodically and then mysteriously re-appear without us being able to discover where he'd been. After he'd performed this magic act a number of times, we began to get wise to some of his secret hide-outs. On a boat the size of *Porpoise* there were lots of spaces he could be holed up in—under beds, inside cupboards, or behind shelves. If we made the effort to search, we would eventually find him in some dark recess, his luminescent eyes glowing in the light of a flashlight. Many times we opened a clothes lock-er to find Miyagi comfortably draped among the stuff, wearing a smirky smile and obviously pleased with himself. That was preferable to the times he would choose to suddenly leap out of his new hidey hole like a jack-in-the-box, jangling our nerves. These sorts of shenanigans may give the impression that he actually enjoyed boat life, but, in truth, he carried an underlying grudge against fate or God or whomever a cat lays blame on that he was taken from his secure home and banished to this uncertain sea life.

So I wasn't overly worried when Lyza reported the cat missing. Knowing how he hated the motor noise, I figured he had just scrammed somewhere when the engine came on and would reappear as usual.

"Don't worry about it, darlin'. He'll come back out once we're anchored and the motor is off."

Lyza looked at me dubiously. "I don't know. I think we better phone the Blaine Marina in case he's on the dock somewhere. Remember that time he end-ed up on Barry's boat."

I did remember the Barry Boat Caper when Mr. Miyagi actually did jump ship, failing to return. Barry was, like us, a resident of White Rock who had acquired a Formosa, a boat with a similar design to ours. He had a berth four slips down from ours at the Blaine Marina. In the early days of our first summer on *Porpoise*, we shifted ourselves and the cat from home to the boat, remaining tied to the dock for a few days in order to do various maintenance projects. After a couple days of establishing his security in the new surroundings, Miyagi got brave enough to explore around the docks. We learned not to worry about these forays at night because he always was back by morning, curled up in his favourite perch on the helm seat in the cockpit. Except for one morning—no

37

Cast Off Your Bow Lines

Miyagi anywhere. After no sign of him all day, we sounded an alarm in the marina community. Word of the missing Miyagi got around to everyone, including Barry at home in White Rock, but no one had seen him. After two days of this, we reluctantly had to leave him wherever he was in order to keep an afternoon sailing rendezvous with friends. We'd just cleared the marina entrance, heading into Semiahmoo Bay, when we got the call.

"I found your cat." It was Barry.

"Oh, wonderful! Thank you. Where did you find him?"

"Not so wonderful. He has been in my boat this whole time. Your dumb cat peed on one of my best shirts!"

I was so excited he had found our lost shipmate that I really had to school my voice to sound appropriately contrite. "I'm so sorry about that, Barry. We'll get it cleaned or replace it for sure. But how did he end up in your boat?"

"Through the overhead hatch. It was the only thing open. Once he jumped down the hatch, there was absolutely no way out. He must have mixed up my boat with yours."

"Sounds like your boat is the perfect cat trap, Barry. Hey, I *am* sorry about your shirt. The poor guy must have really had to go by then."

Barry sounded gruff, but he was only mildly perturbed and also a little amused. Thankfully, he was also heading out in his boat so we did a Semiahmoo Bay cat transfer that could have set off some border alarms if noticed. Getting his shirt peed on was bad enough, but getting arrested for cat smuggling would definitely not have made Barry's day. Miyagi, none the worse for his adventure, was glad to be back on his own familiar vessel. We were hoping that was the end of his cat capers.

So with this latest disappearance, it wasn't beyond the realm of possibility that Miyagi had somehow been left behind in Blaine and was now skanking about the docks oblivious to the fact that his ship had set sail. Our guests were concerned about our missing mascot as we dropped anchor at Sidney Spit and offered to shorten their trip to go back and find him. I was still a little irritated at what a nuisance he was and was more inclined to let things run their course.

"Oh, I'm sure if he's at the marina someone will feel sorry for him and take pity on him," I said. (I knew he wouldn't handle very well missing his three squares a day with snacks in between and that he was totally incapable of foraging for himself).

On the other hand, Lyza was generally more interventionist.

Miyagi in the Belly of the Whale

"We've got to find him! We've got to go back. We can't just leave him to fend for himself." My cavalier attitude wasn't going down well with her or with our compassionate guests, and I began to feel cornered.

"Listen, he could still be on board somewhere, although it is strange that the engine's been off for a couple of hours now and he hasn't appeared. At any rate, we're not going to travel in the dark all the way back to Blaine in search of a cat. We'll just have to wait until morning and decide what to do then."

They all agreed to that plan, and after an enjoyable evening of sunset, wine, and dinner conversation, we all retired. Life on the water, with all the motion and fresh air, has a narcotic effect. I lay sleepily in our captain's berth as darkness enfolded the night, feeling a little worried about our missing mascot. I had to admit the affection I felt for the little guy. This life was proving to be more arduous for our pet than we had anticipated, and now he might really be in some trouble. Thinking back to the beginning, I realized that even his introduction to life at sea had been traumatic.

He had basically been conscripted aboard *Porpoise* that spring. We couldn't leave him home because we had our house listed as a vacation rental when we were away. So with no consultation with him, we decided he should become a sea cat. The transition had been rough. Grabbed one day by the scruff of the neck and shoved into an undersized cat carrier, he was walked, howling all the way, to the end of White Rock pier. Balancing the carrier on the bow of our little inflatable, we zoomed across the bumpy water to *Porpoise*, anchored off the breakwater. Then he was plopped unceremoniously on the deck along with piles of groceries and luggage. He sat there for some time contemplating this alien environment through the bars of his carrier, his landbound cat brain trying to make sense of the strange sounds and smells as his body experienced the sensation of the "ground" heaving beneath his paws. His world was definitely rocked!

Once everything else was stowed, we opened the door of his cage to let him out. As most cats behave in new and strange surroundings, he was spooked at first, frozen in place, staring with wild eyes. Then he flattened his body for a stealthy crawl along the deck, bolting for the nearest cover. Cramming himself between the deck box and the main mast, he let out occasional yowls of protest.

We knew the ways of cats and thought it best to leave him alone for a while to internalize his plight. Eventually he crept out of his cover and cautiously reconnoitered the ground. By that first evening aboard, Miyagi had investigated the whole ship from stem to stern. This exploration even included a sortie out to the very tip of the bowsprit, where he determined there was no way off this

39

Cast Off Your Bow Lines

strange platform and discovered that on every side he was totally surrounded by his least favourite substance—water. Peering down into those dark green depths, he looked like a condemned prisoner in a moat-surrounded jail. I realize I'm making some pretty big assumptions about what a cat's thought processes are, but I bet I'm not far off.

Upon further investigations, he did discover there were a few consolations. Down below decks he found food, warmth, cozy soft beds, secure hiding places, and a defined world where the boundaries were clear and acceptable to his territorial demands. However, just as he was beginning to relax in this new situation, out of nowhere a deep rumbling growl suddenly thundered up from the belly of the boat. His eyes flashed green lightning and his claws scratched madly on the teak deck as he scrambled to get away from this new terror. No safety down below—that seemed to be the source of the roar, so he shot out the companionway and dove for refuge in the binnacle, glowering at me as I steered the ship to the next destination where the sound finally ceased.

He soon discovered nighttime was the best time, as it was almost always quiet and safe. Then he would sit up in the dark, his senses at their peak, listening to the strange sounds of the ocean and its night creatures. Of course his respite was short-lived, because when the sun came up, the roaring, rocking ordeal would begin again. As the days of his incarceration stretched on, we became increasingly aware that he was not making the necessary internal adjustments.

We could almost hear the thoughts running through his head.

How could my life have been subjected to such a rotten turn of events? What have I done to deserve this? What crime? Is there a God ruling over the affairs of cats or not? Why should I be the one consigned to this fate? What possessed my people to embark upon such a weird and stupid existence, and why did they have to include me? I would have been perfectly willing to stay home and watch over the house for them.

It was evident that the usual sunny disposition of our fun-loving feline was giving way to a noticeable melancholy. In the quiet times, we occasionally saw glimpses of playful Miyagi, but the cloud of fear always remained in the atmosphere. With this deterioration of his attitude and in the face of the daily assault on his nerves, we had to face the facts—we had become the guardians of a sour puss.

After reaching the quiet of our anchorage at Sidney Spit with our guests, we fully expected to see Miyagi reappear out of some cupboard or locker. It was late at night by the time we wrapped up our evening conversation. The motor had been off for six or seven hours and still no cat. Lying in my berth, I remonstrated with myself that I should have checked to make sure he was onboard before

Miyagi in the Belly of the Whale

leaving the dock. In the bigger picture, I wondered if it had been a mistake to bring a full-grown cat on a boat and expect him to adapt to such an alien lifestyle. I must have finally drifted off, because suddenly I startled out of a sound sleep. Someone was tapping softly on the door of our stateroom.

"John? Lyza? I think I heard a meow up in the front of the boat. It sounds faint … kind of far away."

Clearing the fog of sleep from our minds, it was surprising how quickly we got to the front stateroom and gathered low over the spot where our guest had heard the sound. I strained my ears to listen. Sure enough, a distant, plaintive meow wafted up from somewhere below. I pulled up the floorboards. It was dry down there but covered with years of layered dust, dirt, and grime. Putting my face on the floor, I shone my flashlight into the bilge, peering up into the gloom. There, at the furthest reach of the bow, was a pair of glowing amber eyes. Miyagi emitted a feeble meow and crawled cautiously toward the light. I grabbed him by the scruff of his neck and hauled him out of that nasty place. Up from the bilge he arose with a mighty smell of diesel in our nose, and I held him at arm's length, avoiding close contact with that filthy ball of fur. There were shouts of welcome, cries of pity at his deplorable state, and a lot of relieved laughter between us all. After closely inspecting him, we found him unscathed, although he would be unwelcome in decent company until he could be cleaned up.

Indeed, the next morning a bath was the first order of business, and poor Mr. Miyagi discovered that the water he feared most had come upon him. Gently lowered into a dishpan of warm water, he was soaked through to his quivering skin and lathered to his twitching ears. He put up with this stoically, but his face was a study of introspection. After this indignity of scrubbing and rubbing until his skin tingled and his fur smelled like perfume, he lay on the pilothouse bench like a sphinx, his eyes alternately closing and opening to little slits. He had the look of one who had gone down into the darkest depths, met the Red Thunder Demon face-to-face, and lived to tell about it. Like Jonah, he had survived being in the belly of the whale.

There he basically remained for two days, occasionally nibbling a bit of food or using his cat box but returning to renew his vigil of silent contemplation. I thought the wheels were turning inside his head, but one can't be sure with cats. Perhaps he was coming to terms with his bad attitudes, or maybe he was mulling over a final mutinous solution. One thing was sure—he had come into direct contact with the two things he feared and hated most in this life at sea. Sometimes the best cure for a fear is to be cast directly into its way, feel it up close, and meet it

41

Cast Off Your Bow Lines

head on. Miyagi had met the shrieking steel and survived an immersion in the water. Now that he had lived through it, there just might be a way to live beyond it.

While he was no doubt lost in such existential contemplations, I did some thinking myself about what must have occurred the previous morning to bring him to such a pass in the bilge. I imagined what must have happened as Dwight and I were conversing over the engine in Blaine. When we weren't looking, Miyagi must have jumped down into the cavity to explore. Then we slid the cover over the engine and the cat at the same time. Once the covers closed over him, he was trapped in the underworld of passageways full of wires, water and fuel lines, and myriad pieces of equipment living under the floorboards that ran the full length of the boat. At first this would not be disturbing, as he generally liked being closed up in tight secret spaces. It would be an opportunity to explore.

Unfortunately, in this dark secret place he was not really alone. Inches from his sensitive ears, the cold steel engine block slept, awaiting only a spark from the ignition key above to suddenly burst into life. As he nosed about down there in the cool quiet, I can only imagine the shattering terror of that instant when I turned the key. He probably jumped straight up with a jolt, crashing his head into the engine covers, only to fall back into the bilge. Scrambling to shove his head into any aperture that could help him escape the shrieking banshee, he would have scratched his way forward until he finally butted up against the bottom of the chain locker at the bow. There he would have stayed, cramming himself into the most compact version of himself he could make. He must have stayed there frozen in fear while the engine roared, and for many long hours after the monster settled down, he stayed and slept until he finally worked up the nerve to let us know where he was trapped.

It took a couple of days to regain his equilibrium. He moved about the boat to different resting places, ate more food, and gingerly walked again the perimeters of the boat. Our friends had made their goodbyes and we sailed off into our summer boat life. On the third—yes, the third—day after Miyagi's ordeal, I first noticed the change while I was nonchalantly peering out the porthole of the aft head that morning. Reaching up to close the latch, my hand came suddenly under attack. A furry paw flashed through the opening, batted my fingers, then shot away. I knew in an instant what must be done. I grabbed at the paw with my right hand while plunging my left through the porthole, where I found and tousled the furry head hidden round the corner. I narrowly avoided a serious scratch.

The claw withdrew only to dart back around from behind the edge of the porthole again, grasping at my hand. I thought to myself, *This means war!* I knew

42

Miyagi in the Belly of the Whale

the game well and played my part, batting back at him with a feint of coun-terattack and ducking down to get a better glimpse of my adversary. There he was—crouched, ready to spring, tail lashing, and eyes glazed over with madness. In this half-crazed state, I knew there would be no quarter given or taken. Miyagi was in a state of high doh, usually only witnessed at home in occasional spurts of prodigious nocturnal energy. Now, in broad daylight, this night raider had come to wreak havoc and challenge the captain himself to a duel. With a bound and a scrape of toenails, he was gone up and over the aft cabin, catapulting over the top of the pilothouse to the front deck where he landed with a clatter, sending the bailing bucket flying into the scuppers. Gaining traction again with his nails, I heard him skittering off toward the bowsprit. After a short silence, I heard a distant thundering gallop growing closer and louder until a blur of tawny yellow shot past my porthole and beyond. A prolonged silence ensued. Thinking the fit had probably passed, I poked an experimental finger out the porthole and wig-gled it provocatively. I barely escaped the flurry of paw and claw as I whipped my hand back inside. Miyagi was back in fine fettle.

The following days confirmed the miracle cure to be lasting. He gamboled over the length and breadth of the ship, perching on various high points in a leonine pose, behaving as though *Porpoise* was his personal kingdom. Balancing along the edge of the railings, he would glance down at the water with bravado, as though it was no big deal. Sometimes, when on a real tear, he even shinnied up the masts. The proprietary way he settled on every bench, bed, and cushion let it be known that *Porpoise* was now duly conquered and considered a part of the dominions of Prince Miyagi.

I thought to myself after witnessing this metamorphosis: *How is it that cats and people seem sometimes to do better after going through ordeals or near-death experiences?* Miyagi's introduction to a Pacific encounter had not been good. But after this traumatic experience, he actually became a pretty solid sea cat, our chosen mascot for the summers. The three of us became known as "the three amigos" in our new vagabond life together. Although keeping up the pretense of independence associated with his kind, Meegs became a sociable shipmate: following us around, planting himself wherever we were, whether at the table for meals, watching a sunset in our chairs at the bow, or in the late evening down below in the salon watching a DVD. When charter guests came aboard, he al-ways preferred the company of pretty girls, schmoozing his way onto their laps for a cuddle. Now *Porpoise* is one of his favourite places to be. Once in a while the old sour puss returns under adverse weather conditions, and he's never really

43

Cast Off Your Bow Lines

happy when the old red monster roars, but he has found a way to be reconciled to his lot, and he makes the best of it. He knows that *"weeping may endure for a night, but joy cometh in the morning"* (Psalm 30:5b, KJV). The sun will go down to bring the quiet of night, and it will come up again with the promise of interesting birds and fresh fish.

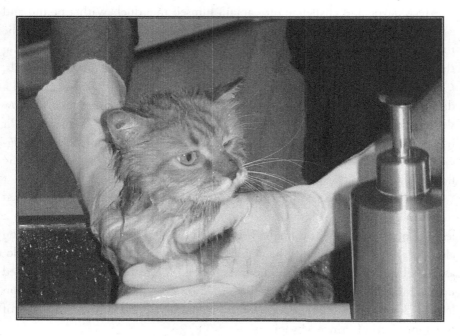

Mr. Miyagi after escaping the belly of the whale

Miyagi in the Belly of the Whale

PPSS

Then Jonah prayed unto the Lord his God out of the fish's belly, And said, I cried by reason of mine affliction unto the Lord, and he heard me; out of the belly of hell cried I, and thou heardest my voice.

For thou hadst cast me into the deep, in the midst of the seas; and the floods compassed me about: all thy billows and thy waves passed over me ...

... And the Lord spake unto the fish, and it vomited out Jonah upon the dry land.

—Jonah 2:1–3, 10, KJV

Mr. Miyagi, proprietor of *Porpoise*

6. Anchoring Can Be a Drag

How in the world did we end up here? I thought as the helplessness of our plight sunk in. No one ever plans to get into a ridiculous and dangerous predicament, but we most assuredly were in one now, with no idea how to get out of it.

Gripping the wheel, I was perched in the helmsman's seat directly above the shuddering wash of the big maxi prop, stern to the wind and revved up to full reverse against gale force winds as they funneled out of the little bay where we thought we'd found safe anchorage. Little by little we were losing ground as our diesel engine strained with all its might against the wind, and *Porpoise*'s high bow pointed like a weathervane toward the expanse of endless Pacific swells spreading away to the north of us. Having just begun to experience the open ocean, I didn't relish the idea of being dragged out to rock and roll among those trackless white caps. Although I dreaded that prospect, at least the day was bright and sunny, moderating the chest-constricting anxiety I was feeling. My immediate reaction in any dangerous crisis is tight-lipped introversion and intense focus on the task at hand, as evidenced by my few clipped words to Lyza when we discovered we were dragging anchor. This pattern of response isn't very helpful to my mate, who craves complete, clear, and verbal direction in such circumstances. Nevertheless, we managed a passable kind of experienced teamwork as together we faced this seemingly inescapable standoff with nature's capricious forces.

The probability of being blown out to sea wasn't the worst of our situation. Thirty minutes earlier, I had poked my head above the companionway to check on our position, and I'd discovered *Porpoise* was moving. She was dragging anchor out of the little cove and directly toward another boat anchored like a sentinel between us and the expanse beyond.

Leaping below to start the engine, I alerted Lyza to our situation, and we both rushed above—Lyza to the helm, and me to the bow to retrieve the slipping anchor. The wind caught the bow and without the steadying anchor to hold us,

Anchoring Can Be a Drag

we spun sideways, moving at an even faster clip toward the boat in our path. Lyza threw the engine into hard reverse as the boat swung round in order to slow our momentum while I kept rolling in the chain. Soon, however, the windlass began labouring under a strain that became worse and worse.

If an anchor is stuck on bottom, the winch will come to a sudden, squealing stop with the gears turning against the brake drum. But in this case, the chain kept moving up in a struggling way, indicating something heavy must be coming up with the anchor. This puzzled me until finally I spotted a large purplish-brown mass suspended a few feet below the surface like a mammoth-sized jellyfish with tentacle-like appendages waving in the current. Glancing up quickly to see where we were relative to the other boats in the bay, I was surprised to see the vessel directly in front of us further off than when we first approached it. It appeared as though that boat was drifting out of the cove at the same rate we were. As I was trying to make sense of this, I glanced back down and suddenly it all became horribly clear. Streaming out toward the bow of the other boat were the unmistakable grey links of our neighbor's chain strung from his bow to ours and disappearing beneath the large blob of kelp that looked so uncannily like a jellyfish. Clearing away some of the mess with the pike pole, I could discern our anchor and then—there was the other anchor wedged into ours. Now it was clear why the other boat appeared to be moving along with us. It was. We were now both adrift, and the only thing holding us both back from heading straight out to sea was our trusty Ford Lehman engine.

As both boats were still slowly moving, even as we had our engine in reverse, I shouted to Lyza to give it more reverse power. But this move caused the boats to start shifting position, us swinging to port and the other ship moving around us to starboard. As the other boat started to spin around us, I could see the anchor chains five feet under the clear water pulling under tension beneath our hull toward our churning prop. I yelled to Lyza to throw the gears into neutral before the chain could hit the prop and disable us, and I cringed as it passed right under the keel and out the other side as the large sailboat tied to us continued to swing like a tethered steer around our stern. With the boats completing a 360-degree circle, we came back to our starting positions still nose to nose, a couple of hundred yards further out of the cove. Miraculously, we hadn't hit any of the other boats anchored in the small cove during this careening cartwheel. We were able to stabilize both boats as our motor roared, rumbled, and grumbled its 120 horses in full reverse. We seemed to be holding our own, locked in this stalemate like two dogs in a tug of war over a rag. With every strong gust we lost a few feet of ground.

47

Cast Off Your Bow Lines

The owners of the other boat had gone ashore to explore this remote beach on the utter north end of Vancouver Island. They were somewhere beyond the call of our voices or the frantic blast of our horn. I dared not reduce the tension on our entangled chains for fear of the unmanned boat blowing away into the trackless Pacific. I wouldn't want to face the owners, whoever they were, upon returning from their little hike. I imagined the scene unfolding ...

"Uh, where's our boat?"

"I think it kind of went that way."

"What happened to it? It can't have just taken off. We have a very good anchor."

"Well you see ..."

My imagination stopped there, as I had trouble even thinking about trying to explain what had happened while pointing vaguely toward the northwest and knowing it was essentially our fault their boat was gone. The ludicrousness of it all would just be too much. Of course, we would also have been honour bound to head out into that gale-tossed ocean to find their boat, with darkness already coming on. No, I couldn't face any of those options, so there was nothing to do but remain in rumbling reverse and hope my nerves, fuel, and transmission could hold out long enough for them to mosey back from their beachcombing.

The morning hadn't started out with any hint of such dire trouble coming our way. In fact, it started out with a bright sun and high spirits after hearing, much to our relief, a benign weather forecast. This was the beginning of our summer adventure to circumnavigate Vancouver Island, and we were holed up in Bull Harbour awaiting a chance to run around the north end of the island past infamous Cape Scott en route to Winter Harbour in Quatsino Sound. Our plan was to spend six weeks exploring the five different sounds located on the wild west side of Vancouver Island, eventually swinging back around the bottom of the island, past Victoria, to return to the more familiar protected inland waters of the Gulf Islands. Our kids were going to take turns as crew, and on this leg we had aboard our daughter, Julie, her husband, Chris, and their son, Kai. Three other sailboats were anchored in the harbour. We had met the crews of a couple of them the day before, and early the next morning, all four sailboats, ours at the end of the line, filed out of the narrow entrance of Bull Harbour into Goletas Channel heading north to run the gauntlet.

The "gauntlet" referred to crossing the notorious Nahwitti Bar, a shallow shoal area that was necessary to navigate in order to get around the top of Vancouver Island. Nahwitti was where five-knot ebb tides met with the prevailing

48

Anchoring Can Be a Drag

westerlies and North Pacific swells, producing sharp breaking seas that even locals feared to transit at the wrong times. As we passed gingerly out of Bull Harbour into the channel, we were pleasantly met with blue skies and glassy seas. Progressing in a westerly direction and riding the ebbing tide, we reached the bar to discover only a slight swell. This caused us to hope for a free pass around the north end of the island to Cape Scott and beyond. Assumptions at sea should never be made.

A few hours into our passage, as we listened in to the radio chatter between our fellow travellers, sombre news reached our ears.

"Have you heard the weather reports for this afternoon? They're predicting gale force winds off Cape Scott."

"Yeah, yeah …well, it's all good … it's all good," the other boat's captain said reassuringly with a Dutch-sounding accent. These two boats appeared to be travelling buddies, and the first one wasn't so sure it was all good. After a pause, the Dutchman said, "Well, we could turn aside and wait it out if you want. I know of a little bight called Fisherman's Bay we could hole up in."

This talk was disconcerting. There had been no prediction of gales on the morning marine weather channel report. Could we have missed such important information, or could the conditions have changed that fast in this wild and woolly place? Suddenly, we felt vulnerable. There was no heading back to Goletas Channel against the tide, and we sure didn't relish facing current against gale conditions out on Cape Scott.

We watched as one after another the sails in front of us turned to port, and we decided to follow the crowd into their perceived safe place. Arriving a half hour after the other boats, we found them anchored snugly in this so-called Fishermen's Bay. It was really just an indentation in the rocky coastline giving protection from the south but totally exposed to the open ocean north. The best spots were taken, but we found a place to drop the hook ahead of the other boats where it was shallower but held promise of decent holding ground. The sun was still shining, and the winds only whispered of stronger forces to come. How could we know the seabed where our anchor lay was strewn with broad garlands of bottom kelp?

Julie and Chris wanted to take advantage of this unexpected opportunity to explore a stretch of coast rarely visited by people, so they motored ashore in the dinghy with Kai to explore the sandy beach while Lyza and I stayed aboard on anchor watch. Soon after their departure, the southerly breeze stiffened over the low headlands, and *Porpoise* strained a little at the chain as her bow headed up

Cast Off Your Bow Lines

toward the beach. The anchor appeared to be holding well, so we went below for a snack to the sounds of the wind whistling in the shrouds. Thirty minutes later, aroused perhaps by a premonition, I poked my head above the companionway to discover that *Porpoise* was moving. Our ordeal had begun, ending in the desperate situation of us holding on for dear life to the other boat and wondering from whence would come deliverance.

To be unexpectedly and unwillingly torn from moorings one thinks are secure, and then to be dragged inexorably toward what looks like the abyss, struck some familiar chords in my psyche. How quickly the circumstances of life can turn against us. How tenuous our vocations, our relationships, our positions in society, our loved ones, and even our very lives. This sudden slide out of control into chaos sparked those same feelings of helplessness and abandonment while fighting against overwhelming forces beyond our control. I could only grip the helm with white knuckles and hold on with shaky resolve, hoping for some intervention from beyond. Of course, such feelings were subliminal, but we soon discovered that, as in those threatening circumstances in life, there is almost always a way out, an avenue of escape, a deliverance unperceived until it suddenly appears, often in a way we don't expect.

Suddenly, charging over the horizon came our daughter Julie at the helm of our inflatable dingy, bouncing over the waves, her long blonde hair flying in the wind. Crouched in the bow was the Dutch couple, clinging to the ropes as they made a beeline for their beleaguered boat. I wouldn't have been surprised to hear the rousing strains of the *William Tell Overture* somewhere in the background as Julie saved the day and dropped the missing skippers back to their ship. We learned later they had heard our horn blaring from the next bay down the beach and were running back when Julie, seeing the emergency, invited them to jump in our motorized dingy and speedily ferried them to their unmanned sailboat. At last something could be done about our helpless situation. I signaled for them to motor up on us to take the tension off the chain, and I prayed it would be enough for me to untangle the mess of kelp and release our anchors.

During all this fracas, another boat anchored in the bay was on their VHF relating a blow by blow to the coast guard in Port Hardy, who usually hope you can solve your own problems before they send an asset out. The coast guard station was a good forty miles away. The couple on this boat, who later became good friends and fellow travellers around the island, offered to throw us a line to arrest our slide. We declined, not wanting to draw them into the entanglement.

Anchoring Can Be a Drag

I could just visualize three sailboats in a tangled mess of rope and anchor chain moving out into the ocean.

Hooking the pike pole under the flukes of the other anchor as they relieved the tension, I gave a Herculean heave and managed to dislodge the anchors from each other. Their anchor dropped immediately to the bottom and arrested their drift outward as we blew past them in the stiff wind and motored further beyond to clear the debris and get our hook on board so we could attempt dropping it in a better place than our original spot. We tried first to get a good hold in the head of the cove but to no avail. We quickly dragged again. Obviously, the bottom was too littered with seaweed and bedded with rock. Where could we go with darkness coming on against this inhospitable coastline? We felt as though now we were the castaways after all.

Just to the south of us along that windswept coast was a long crescent of sand that, according to the chart, had a shelf with a depth of thirty to forty feet a good distance out from the shore, giving us lots of room to drag and no boats to crash into. On the other hand, that strand looked quite open and unprotected as the wind whistled over the low-lying land and raised small white caps even close to shore. Good chance the bottom could be sand as the visible beach was sandy and, with plenty of scope paid out, we just might be able to secure our ship. There didn't appear to be other options, so we prepared ourselves for a possibly rough night.

We pulled up close to the beach and let the anchor go in thirty feet of water. The wind immediately caught the bow, and *Porpoise* careened backwards, reeling out over two hundred feet of chain as we were carried along in the blow, fingers crossed. Gradually we could feel the resistance of the rode over the bottom, and, with a straining creak and scrape of the chain on the bobstay, *Porpoise* swung her bow gracefully up into the gale and tucked her stern neatly behind. The anchor had apparently made a successful grab into the ocean floor, and we were holding.

After thirty minutes of remaining stationary, we began to relax and feel confident about the night. As it happened, the gale blew itself out in relatively short order, and by morning we were bobbing about in an almost flat calm wondering how we could have been so rattled the night before. The other boats were still anchored peacefully in the now notorious nook next to us. In the placid and sunny conditions, it was hard to imagine that the drama of the previous evening had happened at all. So the dramas of our lives often evaporate like mist in the morning sun as we continue on our journey.

Cast Off Your Bow Lines

PPSS

Then they cried out to the Lord in their trouble,
and he brought them out of their distress.
He stilled the storm to a whisper;
the waves of the sea were hushed.
They were glad when it grew calm,
and he guided them to their desired haven.

—Psalm 107:28–30, NIV

Anchoring Can Be a Drag

Grandson Kai's first fish (all by himself)

7. The Charter from Hell

While *Porpoise* doggedly pushed northwestward under autopilot, Lyza and I stood on the bow in a cold wind that blew out of an inlet to the north of us. We would have retreated to the warmth of the pilothouse except for our infatuation with the Dall's porpoises shooting and churning like torpedoes under our bowsprit. They huffed and chuffed about us for five minutes and then veered off toward Gribbell Island, leaving us to feel abandoned, with only vast wilderness and expanses of water surrounding us. We turned out of Fraser Reach into McKay Reach, entering the last stretch of our long journey north to Hartley Bay where we were scheduled to pick up our Scandinavian guests.

We arrived hours later, on September 22, to a high piling dock attended by a few desultory floats with a scattering of fishing and power boats tied to them. Nestled on the hillsides were wood-framed houses clustered along a boardwalk that appeared to serve as the main street. The native village offered little comfort and it seemed as forlorn and deserted as we ourselves felt after making the four-day haul from Port McNeill past Bella Bella through Klemtu and beyond. It seemed as if these far-flung settlements were all that marked the way through the miles and miles of rock, trees, and water we had traversed to reach this particular rendezvous.

We felt a little out of our comfort zone, being up here so late in the year. This northern coast was familiar to us. We had explored it on a few occasions, but never under such a threat of deteriorating weather. Many with local knowledge advised us to get further south behind Vancouver Island by the time Labour Day weekend came around. The North Pacific High would begin its move south toward the end of August, allowing low-pressure systems to travel up from the equator, through the Aleutian Islands, and over this high-pressure zone to continue into the Gulf of Alaska and onto land. As these systems hit the coast, it would seem like the wind and rain never stopped in places like Ketchikan, Alaska, and Prince Rupert, British Columbia, which got nine feet of rain per year.

The Charter from Hell

The settled summer weather was dispensed in shorter and shorter intervals as you travelled north, and the fair weather window for Hartley Bay was long past, if it had ever arrived that year.

A storm was brewing off the coast with winds of over fifty miles per hour, and it was due to arrive the next day. We didn't like storms at the best of times and usually chose to hole up and wait them out. The prospect of a storm coupled with the pressure of guiding and feeding a large group in inhospitable, unfamiliar territory multiplied our anxieties, especially for Lyza who always felt obliged to provide the very best cuisine and comfort she could for our paying customers. When you added to this the fact that our clients were perfect strangers and didn't speak much English, it was a recipe for the perfect storm, both inside the boat and out.

I was much relieved to meet Lars, our bear guide, when we arrived at the dock. He was young, handsome, had an affable disposition, and was the only one of our new crew to speak fluent English. We took an immediate liking to him and felt we would be able to relate to him as the point man. He had contacted us almost nine months earlier to plan this trip, and part of his job was to take clients out to faraway places to shoot bears, not with guns but with cameras. He and his guests had shot polar bears in Svalbard, Norway, as well as Kodiak grizzly bears in Alaska.

Having heard stories about British Columbia's iconic Kermode bears, Lars wanted to build a new itinerary around capturing photos of these unique white bears. Kermodes exhibited a genetic anomaly that made up to 30 per cent of the local bear population white, despite technically being black bears. The natives revered them, calling them "spirit bears," lending them a special mystique.

Kermodes were the main quarry of his clients, who showed up soon after Lars. They were serious customers, decked out in the latest outdoor camo and weighed down with massive packs crammed full of camera gear, complete with bazooka-sized telephoto lenses and an assortment of attachments.

We were introduced to a sweet couple in their late forties who were courteous, blonde, and spoke little English. Next came a sturdy Viking-like woman and her twenty-two-year-old daughter, who seemed to be surveying everything, including us and *Porpoise*, with a certain wary disdain.

These two, I thought instantly, *could be trouble*. First impressions can indeed be prescient, for the first thing the mother did was corner Lyza in the pilothouse to complain about the young bear guide.

55

Cast Off Your Bow Lines

"We are not happy with him," said the mother in heavily-accented English. "He took the best place on the bear-viewing platform to take his pictures. Don't you think a guide should give way to his guests?"

Her furrowed brow and grimace came across as a thunderous portent of things to come.

She crossed her arms and waited, six inches from Lyza's face, for her to agree with the negative assessment of the young man. Lyza hardly knew the guy and quickly looked for a way to extract herself from the clutches of the woman's toxic aura. Excusing herself, she slipped around the woman, who had trapped her in a corner, and shot a wide-eyed warning glance off my bow to signal approaching danger. Shaking off this initial portent, we helped the rest of the group stow their heavy gear on board. After we got them settled in their quarters, we hastened to head out so we could get somewhere safe before nightfall and the arrival of the predicted storm-force winds.

Once under way, we were surprised when Lars informed us that they had already taken pictures of Kermode bears. The indigenous people of Hartley Bay ran a program in which they ferried guests on power boats up to Gribbell Island river basins to photograph the bears. The natives had built platforms to give a comfortable vantage point from which to view the Kermodes as the bears took spawning salmon out of the rivers. *National Geographic* had used this same place and guides while researching spirit bears for an issue we actually had onboard *Porpoise*.

We were taken aback by this discovery, as we'd assumed we would be blazing a fresh trail into the wilderness in search of the bears. It felt a little prosaic to discover the natives had beat us to it and had even built viewing platforms. I suspected that the bears were even on the dole. You know, "How 'bout a deal? We'll leave fish around for you if you give a good pose for the tourists."

Lars explained that he'd made the earlier booking with the guides at Hartley Bay to make sure his clients actually got some pictures before they struck out with us into the more remote inlets and river basins. With us he hoped to find more Kermodes and other kinds of wildlife, including grizzlies and wolves, as we penetrated into the far recesses. We shook off any cynicism we felt and got on with our mission. There was no question we were out in a serious wilderness, and if the natives had figured out a way to capitalize on their strategic location and knowledge, so be it. There was plenty of room for us all to explore and discover the wild riches of the central coast. Quite frankly, it took some of the pressure off, knowing our guests had already seen some of these elusive creatures.

The Charter from Hell

Lars had done some research and studied the charts of the area and was convinced we should go first to Surf Inlet, which penetrated deeply into Princess Royal Island, prime Kermode country. He had also heard there might be wolves and other wildlife up there. "We could head in that direction and maybe stop somewhere on the way," he said enthusiastically.

"Lars," I said cautiously, "we want to do everything we can to get you where you want to go, but storm-force winds are predicted to reach the coast by tonight, and we need to be anchored somewhere safe,"

He pressed his point regardless. "I've looked at the chart, and it looks like there is a sheltered bay on the way. Could we stop there for the night and go to Surf Inlet the next day?"

I had just met this nice young man and desired to please him, but my better judgement rose against his idea.

"Lars, I know of Barnard Harbour, the one you're talking about, but I think it's too near the coast and right in the path of the storm. We need to get ourselves further inland until this big wind passes."

I hated being in conflict right off the bat. I didn't tell him we'd never been in Barnard Harbour, or that we never wanted to ride out a blow in an untested place. We were to compromise under pressure later in the week and pay a dear price for it.

"I think we should go up Fraser Channel to Butedale, where we can tie up to the old wharf under the mountains. I'm not sure how long this wind will persist tomorrow, but there's a reason Surf Inlet has its name. The mouth of the inlet opens directly west into the waters at the base of Hecate Strait, notorious for generating huge seas. Those seas will roll right up that channel into our anchorage, exposing us to wind and swells and making it very hard for us to get back out again."

I didn't tell him I'd never been in Surf Inlet either, but I hoped the name would cool his ardour for it. It sure did mine. I needed to talk him into seeking refuge in the mountainous inlets to avoid places nearer to ocean exposure. I didn't want to mention that this was not the only storm predicted for this unfortunately-timed week of his charter. No need to pour cold water over the whole trip. Besides, you never know how things will go. It might not end up as bad as it sounded on the weather channel.

"Yeah, okay. You're the captain. We'll trust your judgement."

"Thank you. My first responsibility is to make sure you're safe."

Cast Off Your Bow Lines

I could tell he was disappointed, being young and fearless, but I also knew he didn't fully appreciate the dangers. So off we chugged into the sodden grey where sky and sea blended into one and the shorelines and tall mountains were indistinct. We left in the afternoon, and it was at least twenty-six miles to Butedale in the darkening September light. Hoping my judgment *was* right but feeling badly about diverting from his chosen itinerary, I couldn't help having a sense of dread over the whole venture. We were skirting the edges of our comfort zone by hosting a Kermode bear charter late in the season on the inhospitable north coast. It felt like we were heading into waters above our heads. As the rain persistently increased, so did my apprehensions on many levels.

I worried about water penetration below decks. *Porpoise* could be cozy and comfortable on rainy days. Our Dickenson oil stove located in the salon radiated a steady warmth, keeping guests, clothing, and bedding dry and spirits high; however, a constant onslaught of rain soaking decks, people, and clothing eventually could overwhelm our defences. Our ketch was a tight ship, but she could become a little incontinent in certain places under severe conditions of horizontal rain due to strong winds. We were used to a lot of sunny summer excursions behind Vancouver Island in our chartering life where managing the occasional rain was easy.

The other deep source of apprehension was the challenge of finding safe places to anchor or tie up in storm-force winds. Our guests wanted to travel into the far-flung reaches of the inlets in this wilderness where we had not tested the anchoring grounds. There are pressures enough without adding an insecure place to spend the night. Butedale was at least familiar to us and had an old float to tie up to. The thought to trying to reach Surf Inlet some forty miles west toward the ocean was unthinkable to me at that point.

We motored along in the drizzle for four-and-a-half long hours, settling our guests in their quarters and becoming acquainted over tentative conversation, hot tea, and cookies. Lyza started to crank out her cuisine, the best defensive strategy against bad weather. A successful charter is all about good food. Meanwhile, I tried to paint a positive picture of our destination, a bit of a tough sell as Butedale was only a derelict old ghost town. As it finally hove into sight around the last bend, it was hard to put a good face on the lonely desolation of collapsed buildings slumping into the sea. I could only hope our visitors would find it a quaint example of British Columbia's logging and fishing history.

Butedale was founded in 1918 as a fishing, mining, and logging camp that hosted a salmon cannery and a community of four hundred people into the early

The Charter from Hell

1950s. Its most distinctive feature was a large waterfall draining a big lake behind the town. The inhabitants installed a turbine below the falls that produced abundant electrical power for the whole area. For years after the place was abandoned, the power was left running, lighting up the buildings and streets. Boats passing by in the channel at night would see the whole shoreline lit up, with streetlights casting an eerie light over empty buildings and streets.

Those spooky days were gone now, and very few lights greeted us when we swung up to the float. It was a dismal affair, the dock consisting of ancient water-logged cedars floating low in the water and held together by rusting cables and rough grey planks rotted through in places. Not an auspicious beginning, I thought, as I gazed up at sagging structures that hung precariously over the water. But I was grateful for the haven. Outbuildings and old bunkhouses were scattered about the rocky shoreline, half of them tumbled into themselves and filled with broken timbers and relics of a past bustling existence. It elicited a sombre sense of veneration, almost as though we were visiting a cemetery, marking past lives with tombstones made of old posts and remnants of rusted machinery poking out from the grass here and there among the ruins.

Vegetation and trees marked sixty years of the wilderness taking back its claim on the town. Thickets of scrub brush, berry bushes, and salal invaded doorways, covered windowsills, and overhung dilapidated sheds. Large evergreens grew in the middle of old streets and other unlikely places, revealing how long it had been since the place was inhabited. Butedale was only one of a number of such ghost towns on the central coast to have suffered the same fate. Places like Namu, Ocean Falls, and smaller outposts like Port Allison spawned at the turn of the twentieth century by a fishing, logging, and canning industry that had run their course in the boom times and were now abandoned. Without the means to sustain people in this country, accessed only by boat or float plane, this territory was now much less populated than seventy to a hundred years earlier.

With the passing of these foundational industries, tourism became the only promising economic activity. Old landmarks, such as this one, invited visiting hikers, kayakers, and cruisers to explore the vicinity and contemplate the history of this mystic coast. We hoped the dead little town would cast a spell over our bear photographers, help mollify the weather disappointments, and spur in them an appetite for adventure. As the wind howled that night, I was even more grateful for our moorage, humble as it was.

The wind had blown itself out temporarily by the following morning, and I discovered that our sturdy crew weren't easily discouraged but were quite intrepid.

59

Cast Off Your Bow Lines

Marching out in single file with me at the head as guide, we tramped through the remains of the town to a trail that led up into the surrounding woods. Suddenly we were enclosed in primal rainforest, scrambling over roots and rocks through dense fern, moss, and huckleberry. We hiked up hillocks and down into ravines scented by the all-encompassing evergreen forest of Douglas fir, hemlock, cedar, and pine. The spongy land was already soaked to the full by the early fall rains. Now with the onslaught of these storm-borne downpours, our rough-hewn trail was turned into a muddy stream, fed by a maze of little rivulets running among the moss and roots of trees.

Slogging over the muddy track, we came upon meadows of verdant flora. Turning to look behind in one such clearing, I was amused to find my amiable guests laid out in various contortions on the ground among the wet plants. They were aiming the inquiring eye of their lenses at the many specimens of unique Canadian flora while staying alert for any possible appearances of fauna. They were all clothed in the latest waterproof camo, and I began to feel like I was at the head of some military campaign. The analogy was actually an accurate description of the attitude and aptitude of my photographic expedition. I found some comfort in this fact as I increasingly realized the timbre of these Scandinavian souls. If we had to go through a perilous week of north coast bad weather, at least our guests were of hardy stock and not of the tenderfoot variety we often had further south. They were serious-minded explorers, prepared to meet whatever British Columbia served up. At least that's how it appeared then. We would discover later that not all were.

I had travelled this trail once before and knew there was a lovely lake full of fat trout at the end. Seeking to find silver linings in our cloudy weather, I had packed a couple of small casting rods. My thought was that fresh trout caught by our guests for dinner could mollify some of the disappointment. As we reached the cluster of logs that invariably jam the outflow of these kinds of lakes, I was surprised to find my photographers more interested in traipsing about the shore with their cameras than fishing. Incomprehensible as this was to a trout fanatic like me, I obliged their preference by leaving them to it and spending my time pulling in ten fat trout in short order. Even though they showed no interest in catching the fish themselves, they had no aversion to consuming them later that evening. I began to think maybe this trip could turn out not so bad after all as Lyza and I checked off our first twenty-four-hour day.

We left Butedale that same afternoon to run for Khutze Inlet, our next opportunity to fulfill our guests' desire to encounter wildlife. Khutze was a

The Charter from Hell

promising destination as we plowed through the sheets of rain that returned to accompany us on our journey. The overcast grey sky blended with the grey water and obscured our surroundings, obliterating the mountains and even the tops of the tall evergreens around us. Feeling glum in the gloom, I switched on the coastal weather station to get the latest update. The news wasn't good. The previous night's storm had hit the coast with a vengeance, almost taking out an exclusive Japanese-owned fishing resort built on a barge anchored in Barnard Harbour, the very place I had decided to avoid. I couldn't help but feel vindicated with Lars after hearing the report. The resort had lost three of its four anchors in the blow, provoking a rescue and evacuation of all the guests by a ferry diverted from its route to the lodge's location. My caution had clearly been warranted. I glanced sideways at my young bear guide sitting on the cockpit bench.

"Good thing we weren't anchored there last night, eh, Lars?"

I think I will be forgiven if I felt a touch smug on the inside, although I think I masked it pretty well. To his credit, our bear guide nodded appreciatively. He was beginning to moderate his eager impatience in light of this firsthand experience of British Columbia's capricious north coast conditions. However, one of his clients was beginning to manifest the dark side of her personality in collusion with the inclement weather.

The whole coast from Prince Rupert to Cape Caution was buried in this depression, so there was no attractive destination to head for presently, but Khutze Inlet looked protected from the wind and contained a large river valley, just the kind grizzlies favoured. Although we hadn't explored this body of water before, I'd seen photos taken from its head where a lovely waterfall cascaded down a cliff. Since there was hardly anywhere else to target within the scope of our day's travel time, we felt our way into the narrow entrance.

Around a bend further in, we entered a large bowl of deep water where the only good anchorage depth could be found just off the spectacular falls ... exactly where the only other boat in the whole inlet was already anchored. I sidled up as close and safely as I could and dropped the hook in seventy feet on a steeply shelving bank of what we hoped was river mud. If we swung on our chain toward the falls, it was quickly shallow enough to see bottom. If we swung the other way, our depth finder read 140 feet. It would have to do.

Looking up from the depth finder, the first sight to greet Lyza and me was a large black, almost sinister-looking, inflatable coming toward us from the other anchored boat. The man at the helm wore heavy weather gear and a smile as he

61

Cast Off Your Bow Lines

approached *Porpoise*. As he pulled up alongside, the affable driver struck up a conversation with us.

"Are you here to stay? I heard you talking on the VHF to the other boat that just left here. Sounds like you have some questions about this place."

I was gratified by this helpful offer and filled him in on our situation.

"Yeah, we have some people aboard who are looking to photograph Kermode or grizzly bears. They're photographers from Norway. We weren't sure what we might be getting into here and hoped to gain a little more local knowledge of the place. Our intention had been to explore the outside inlets and river outlets of Princess Royal Island, but this crazy weather has limited our options."

"Oh … interesting. I do the same thing. I bring people out here to see the Great Bear Rainforest." We kibbitzed for a bit, as mariners do, about the calamity of the night before, with the ferry rescue getting top billing in the storm news. He confirmed that it was some of the worst weather for September in over fifty years. He turned out to be a fountain of information and very experienced in these kinds of trips.

His sturdy old power boat, appropriately named *Great Bear II,* seemed to match the hardiness of the massive tender he was standing in during our exchange. I eyed the sixty-horsepower motor on his inflatable, knowing our little tender had only a fifteen horse. The longer he talked, the more I started to feel woefully under-equipped and out of my depth for this sort of place. He talked about the dangers and the attractions of Khutz, and I was encouraged as he pointed across to the bay's far side where we might find some foraging grizzlies on the grassy river delta.

In the course of our talk, we discovered he was a Vancouver firefighter who worked with a close friend of ours. He loved this part of the province so much that every year he bunched all his vacation time into one lump to use it for his rainforest charters. Waving him goodbye, we got down to dinner prep and went to bed that night with the comforting thought that the next day held promise of some genuine wildlife sightings.

We woke to incessant rain—it hadn't stopped all night, and the outlook didn't look promising—but our guests weren't deterred. We headed to the river delta in our tender to find those grizzlies. Carrying five large adults pushed the weight limits, especially with the abundance of camera gear, and I felt increasingly uncomfortable with the situation the further we progressed into the river delta. The channels penetrated deeply into the forest, and as we left the tidal influx, I

62

The Charter from Hell

felt the river push against us. Eventually we were hardly making headway against the outflowing fresh-water current.

Staring over the starboard side, I swerved to avoid an underwater log in our path and noted that we'd only cleared it by a few inches. How easily the strong current could carry us over that log to upend and swamp us in a startling moment. There they all were—squeezed in my boat, clothed in that fancy outdoor wear with boots laced on their feet and thirty-pound packs. I was skirting with a disaster of unthinkable proportions. Safety had to come before their adventure, even if I ended up disappointing them. I turned to Lars, who didn't look worried, and I knew it was up to me to alter the plan.

"You know," I said as casually as I could, "I think we have too many bodies in the boat for this current. I don't want to attempt to get any further upstream."

At that, he seemed to notice we weren't making much headway and volunteered a solution himself.

"How about you drop a couple of us over there on the long grass of the delta? We can search around for bears and you could drop the other two further downstream."

Looking around for grizzly bears on foot in long grass wasn't something I would relish doing, but this was Lars' area of expertise (or at least I hoped it was), and I was relieved at getting that much weight out of our overloaded boat. I dropped the three of them on shore and headed back down the river to where the mother and daughter requested to be let off near a big drift log that lay beached amid the long grass. They'd seen some plants they wanted to photograph.

"Okay," I said, "I'll be back to pick you up in a few minutes. I'll just check on the progress of the others."

I wasn't comfortable having the two groups separated, with me running back and forth on the strongly running river and knowing the tide was on its way up, but I wanted to give the guests the best possible experience doing what they came for. I made it back upriver to fetch the first party, but there was no one on the beach. The group had traipsed off, armed only with a can of bear spray (apparently all he ever carried on his forays). Such bravado was fine for him but scary liability for me if he became bear bait. I was beginning to feel that combining his bear adventures with our sailing adventures was perhaps not the best idea. I understood *my* risks, but I wasn't sure I wanted partnership in *his*.

All three finally showed up after about twenty minutes to report that the bear search had been a bust. *Thank God*, I thought, although I felt a little guilty about being so relieved when they appeared to be so disappointed. I convinced

Cast Off Your Bow Lines

them to get back in the boat, as the tide was quickly rising, and I was worried about our charges back near the mouth of the river. By the time we got back to the ladies on the log, the tide had disconcertingly come up two feet, leaving them to balance on the drift log surrounded by water, with only the tops of the long grass waving above the current. Thoroughly rattled, I helped them back into the inflatable and breathed a sigh of relief as they climbed the boarding ladder back to the security of *Porpoise*.

Later we learned from the *Great Bear* skipper that while he and his guests were motoring up a different stream of the river delta, he had seen two huge grizzlies just yards from where Chris and his people were standing. The idea that such huge carnivores were so close yet unseen freaked me right out. Not the Norwegians. When they heard that, they wanted to get back out and try for a close-up. Thankfully, darkness was falling, so that wasn't going to happen, and we settled in for the night.

Our nightly ritual of listening to the marine forecast for the following day was now part of everybody's schedule, since so much depended on what that monotonous voice said next. The voice confirmed that the next weather event would be bearing down on us by the following evening. Fifty-mile winds with much higher gusts were predicted, accompanied by up to eighty millimetres of rain. As our guests stripped off their dripping wet gear to hang it on the retractable clothesline above the diesel heater in the salon, the interior of the boat felt warm but humidly claustrophobic. That, however, was much better than being left outside in the elements, and we were thankful for our cozy retreat. It was always a source of wonder for me to contemplate the contrast between our little capsule of warmth and light with the reality that we were dangling precariously off the edge of a river delta and buried in the belly of British Columbia's north coast wilderness.

Unfortunately, a certain component of this environment was stealthily making inroads into our redoubt, seeping through little cracks and seams in the deck and making its way toward the bedding of a couple of our guests. As the weather deteriorated, so did the mood of these two crewmates, the mother and daughter. The other crew were stoic and even cheerful in all our weather trials, but the two weak links continued to express discontent. They continually criticized their leader behind his back. I didn't want to know what they might be saying about us or our beloved boat to their leader.

We left early the next morning to head down Graham Reach and Tolmie Channel to Klemtu, another isolated native community. There we hoped to tie

64

The Charter from Hell

up to the dock and find protection from the next blow. We also hoped to distract our guests with the prospect of exploring the little village and its new authentic long house. A far cry from stalking spirit bears in the wild, but at least we'd be safe. I did have to discourage Lars one more time from us heading up to Windy Bay (Wouldn't even that name give him pause?) in the very remote Kynoch Inlet. I knew he was feeling the pressure of time running out on his dream of Kermodies in their natural habitat, and having been to Kynoch ourselves, I could attest it was the penultimate wilderness destination, but we had done it in settled conditions in July!

Arriving midday at Klemtu, we tied up to the small floating dock and surveyed the cluster of modest homes built around an enclosed crescent-shaped bay into which a river flowed from the surrounding mountains. A local elder greeted us at the dock, eager to be our guide through the town and our story-telling host at the long house. After learning some interesting history of the place in the cavernous lodge, and a few tall tales that stretched our credibility, we trekked back to *Porpoise* on the one main road.

Soon I was back to fretting and stewing over our sadly lacking sodden itinerary. Discovering another skipper on an old, interesting-looking power boat tied across from us on the rickety float, I introduced myself and began probing his knowledge of the locality for some alternative to Klemtu for the rest of the day and night.

"I really want to get these folks out to some more remote location with a possibility of finding some wildlife, but it has to be safe from these coming winds. Do you have any recommendations?"

The gangly fellow with greying hair had been a kind of mobile missionary to this area for over a quarter of a century and seemed a likely person to know of such a refuge. He pulled out a beat-up chart from a cupboard in his pilothouse and spread it out on the chart table. After a couple of minutes, he put his finger on Alexander Inlet and tapped on it for emphasis.

"This spot could fit the bill. It's just a few miles up the main channel, so it wouldn't take long to reach, and you could tuck in nicely from the wind at the head. I've seen wolves in there a time or two."

Gazing down at the contours of the water and mountains as they appeared on the chart, I scrutinized the long, snaky inlet marked with small islands and punctuated with a number of Xs marking the many rocks to be avoided. The entrance and ensuing channels were quite narrow and led down around a bend

Cast Off Your Bow Lines

to the end, where a small bowl of a bay appeared to offer snug protection among the surrounding hills.

"That looks like it could be pretty bombproof," I replied, using the boater's term for an anchorage with wind protection from all angles. I looked around me at the little village, where the main entertainment for the locals seemed to be driving their pick-up trucks the half-mile road from one end of the crescent bay to the other. Assessing the mood of my guests now lounging on *Porpoise* after their walk around town, I heard myself mutter, "I've got to do something."

I went back to Lyza and showed her the spot on our chart. She agreed it looked pretty safe, and we made the decision to go. *Life's an adventure*, I thought to myself. We both felt it was better to make the effort than to cower in Klemtu.

So off we chugged back up the channel in the enveloping mists. We soon turned into the entrance of Alexander and threaded our way for eight miles around islands and rocks while the skies darkened and the rain fell with increasing intensity. By the time we reached the little pocket at the end and dropped the anchor in the downpour, we could hardly make out the shoreline through the pilothouse windows. Camera work ashore would be impossible with so much moisture on the lenses. Even our intrepid bear guide didn't seem eager to slog about the uninviting shore. Surely any purported wolves who dwelt here would have the sense, unlike us, to hole up in their dens. I felt defeated. We'd come all this way from Klemtu only to hunker down in incessant rain. We might as well have stayed where we were. The late September darkness fell swiftly around us like a smothering veil and matched my mood exactly.

Nothing remained but to bear it. Lyza's efforts to serve up even more tantalizing dishes to make up for the charter's shortcomings seemed to be edging on desperation to me, and I was a little worried I might end up with a "mutiny on the *Porpy*" if conditions didn't improve soon, but we both rallied our flagging spirits and endeavoured to pass the evening with our guests in pleasant conversation and board games. In the background, tensions lingered among the crew. The twenty-something daughter had sequestered herself away from the rest of us in her bunk all day with a stack of the Scandinavian equivalent of Harlequin romance novels. It was clear she was just enduring this "vacation" and wanted to be somewhere else. It was also clear to me that Lyza shared those sentiments, but she put on a good face with the guests. Soon we all shifted ourselves off to bed.

That night the winds increased, and the predicted depression sent gusts howling down the valley of Alexander Inlet. Rather than protection, the mountains served as a wind tunnel, funneling the southeasterly gales straight down

The Charter from Hell

the inlet, round the bend, and into our supposed enclave of safety. As the wind gusted high and roared in the trees, it caught *Porpoise* on her bow, straining the anchor chain taut first one way and then the other. Down in our stateroom, the most unnerving sound was the loud scrape of the chain against the bobstay cable reverberating through the ship as the bow jockeyed back and forth. But the anchor held its own against the abuse, and we finally dropped off into fitful sleep.

Maybe a subtle change in the cacophony of noise outside caused Lyza to awake around five in the morning. She has always been sensitive to the boat's various noises, and she lay alert in her bunk for a couple of minutes, listening intently. Then she was shaking me awake.

"John … John …wake up!"

Sleeping lightly myself, I instantly heard the anxiety in her voice.

"I don't feel good about what's going on outside. I think you need to go up and check on things. See if we're dragging anchor. It doesn't feel right to me."

I was reluctant to get out of the warm covers to face the cold blowing wind and rain, but I had learned from past experience to trust Lyza's intuitions. It only took a little hesitation before I was donning my clothes and parka and stumbling up the companionway.

I hustled back to the protection of the dodger, where the chart plotter made a glow over the helmsman seat. Squinting at the dim screen, I toggled up the brightness and dialled in a close-up view that showed a dotted line marking the path of the boat over ground. My heart sank at the sight. The dotted line traced across the bay a full three hundred yards from where we'd started the previous evening. That line had taken us out of the safe depths marked in white to the shallower depths marked in blue, with the green depths near at hand. Green was where the bottom went dry at low tide. We had gone from over fifty feet of water below us to less than twenty.

Though alarming, this didn't mean the anchor was loose on the bottom; otherwise, we would have long ago hit the beach. It meant the anchor was dug into the mud but plowing a few feet every time a hard gust of wind hit the bow. At twenty-six tons, our ship put tremendous pressure on the anchor and chain when a hard blow caught our big bow. Chain is stronger than nylon rope and generally better for holding power, but it has no elasticity under pressure. Slowly but surely, a few feet at a time, *Porpoise* was headed for a rendezvous with the rocky beach of the bay, miles and miles from any help or habitation.

I sat and stared at that screen, wishing it to give me a better report. At least we hadn't woken to the thump and bump of our keel crunching on rocks. We

Cast Off Your Bow Lines

still had a little way to go before that impact, judging by the distance on the chart plotter. The little triangle representing our boat wasn't right to the rocks yet. As it was still inky black outside, the only measure of our position was that tiny black triangle on the small square screen. I willed it to stay where it was. But after watching its position for a few minutes, it was clear that the black triangle was moving slowly but inexorably toward the shore. We had to do something.

I ran down to Lyza in our stateroom. Not surprisingly, she was sitting up in bed with her clothes and the light on.

"We're plowing toward shore. We've got to move the boat now. We can't wait for the morning light. We'll just have to pull up the anchor in the dark, motor up into deeper water, and drop the hook again. That will buy us some time until it's light enough to get out of here." This was a lot of explanation from me—I usually get very taciturn in any kind of crisis.

Now it was Lyza's turn to go quiet. She nodded and leapt out of bed, knowing what was required of her. Soon she'd snapped on the deck lights, pulled out the spotlight, and perched herself at the helm. I stood near the bowsprit with pike pole in hand and my foot on the anchor winch button as we began the ticklish job of bringing up the chain as the diesel engine rumbled under our feet.

As a few feet of chain passed over the wildcat and into the locker, a gust would catch the boat causing the bow to swing away until the chain crimped in the winch with a grinding crunch. Then Lyza would rev the boat forward and crank the wheel hard to pull the boat back nose to wind so I could get enough slack to pull up a few more feet of chain. But the momentum of that manoeuvre in the wild wind would cause the bow to swing back too far and then it would fall off in the other direction. In the interval between these mad plunges, I'd scramble to winch in as much chain as possible before the next jarring crimp stopped the movement in that direction.

When the anchor came off bottom at last, I hurriedly reeled in the final thirty feet of chain, because at that point the anchor was no longer a stabilizing factor. We faced the immediate danger of the wind carrying the boat broadside straight back onto the rocks. I heard the engine roar as Lyza gave it full power in forward to gain control and knew she would have the wheel hard over. She was trying to counteract that careening slide toward shore and get the bow back into the wind so we could motor back into deeper water. I glanced up, hoping to see something recognizable as a landmark, but it was still pitch black and I wasted no time bolting for the helm.

The Charter from Hell

I knew how fast even a solid ship like *Porpoise* could slide across the water under a strong gale, and I had no idea how many feet remained before impact. I ducked in behind the dodger, and Lyza slid out from behind the wheel, gladly yielding it to me. With the engine gunned and the helm jammed as far as it could go, I yelled to her to man the searchlight as I desperately tried to orient myself as to where we were. The electronic image on the screen is notoriously inaccurate as to the direction the boat is pointed, and the triangle's proximity to the green shade denoting the shoreline did nothing to assuage my fears.

As she frantically scanned the dark, I whipped my head around the edge of the dodger to get a look. My heart stopped. The light suddenly revealed rock and trees directly off the bow. I was propelling us toward disaster at top speed! My reaction was instinctual—there was no time for thought.

I slammed the transmission with a loud, jarring clank into full reverse. The engine roared and the propeller churned furiously under our stern. We held our breath with shoulders hunched, bracing for the impending crash and the inevitable crunch of our keel on the rocks.

It never came. To this day, I don't know why not. We were almost on the shore—I could have practically touched the overhanging tree boughs—yet as the diesel churned the water, our forward motion slowed, and we reversed into the dark and the deep. *Porpoise* had gracefully managed to bow out of that disastrous meeting. The chart plotter confirmed we were moving into safe water, and this time it proved trustworthy. The emergency and my instincts propelled me to do the thing I should have done in the first place, and since that day I repeat these words to myself as almost a religious dogma: "When in a tight spot under windy conditions, lose the bow and back yourself out!" The boat is easier to control reversing into the wind. It was hard-won wisdom, and it has come in handy on a number of dicey occasions—thank goodness none as dire as that one again (well, maybe one).

Lyza and I were still shaking as we found ourselves safe and floating free. With plenty of sea room, we turned the boat around and motored into the centre of the bay to drop the anchor again. Our hook held, and we felt the reassuring swing as *Porpoise* hunkered up nose into the blustery gale. We had no illusions that we would stay put, but we had bought ourselves enough time for daylight to arrive. We remained on anchor watch until the sky finally began to lighten after an hour or so, revealing a dim outline of mist-drenched evergreens surrounding us on every side. They gradually morphed from dark night terrors to indistinct grey ghosts and finally into friendly green giants welcoming us back into full

Cast Off Your Bow Lines

daylight. The low tide on shore revealed the stony strand we had almost landed on, carpeted now with yellow rockweed that added a golden touch to the scene of green. By morning coffee time, the wind and rain slackened to almost none.

Incredibly, our guests had slept through the whole ordeal.

"Didn't you guys hear the commotion last night?" I queried as the couple emerged from their berths and gathered around the table for the day's first cup of French-pressed dark roast.

With a little translation help from Lars, the pair basically communicated the following: "At one point we heard something going on above, but we figured whatever it was, you guys had it completely under control. We just rolled over and went back to sleep."

I liked this couple and appreciated their matter-of-fact trust in us, even if it had been misplaced. Lyza shot me a glance, but only I would have noticed the almost imperceptible head shake. I agreed it was best to let them abide in the illusion of security than explain what really happened. Better they not know that, instead of sipping coffee at that moment, we could have all been shipwrecked—marooned on a cold British Columbia shore—hoping we could rouse the coast guard on our radio to come and rescue us.

We could hardly wait to get out of that isolated hole. Soon we were wending our way out of desolate Alexander Inlet's reef-infested channels with our tails between our legs. There was no use trying to find some wolves that morning—I didn't even have the stomach to try. This charter was shaping up to be about as big a bust as it could be, and my feelings of dejection only increased as we picked up the radio signal for the weather channel. I was praying for good news but instead received the worst report yet. This time the storm heading our way, the third in under a week, would wallop us with hurricane force winds (above sixty-four miles per hour). This was unbelievable to me.

It's one thing to handle terrible weather when you're out on your own. You can just hole up at a dock in some port until it blows over, turn up the diesel heater, make some tea and chocolate chip cookies, read, and relax. It can be a very pleasant down time. However, when you have people aboard who travelled across the world and paid you the big bucks to experience the wonders of the British Columbia coast, you feel responsible to make something happen. Feeling under that pressure had landed us in a lot of trouble the previous night. Nevertheless, I still felt the pressure to pull something spectacular out of my hat. We really needed something positive to happen to salvage this charter—anything.

The Charter from Hell

When we rounded the point into Tolmie Channel, the skies started brightening, and we saw patches of blue sky ahead of us. Shafts of sunlight broke through, striking the sea and making the water sparkle. Lars stood at the bow, and I figured he was probably struggling with his own disappointments about this trip. Suddenly, he shouted something unintelligible and pointed off the port bow. We could just make out that he was saying.

"I think there may be whales ahead of us in the channel!"

Lyza didn't miss a beat. She grabbled her special binoculars and was at the front of the boat scanning the water before the rest of the crew started tumbling out of the companionway. She's a whale nut ... actually, any kind of cetacean, but humpbacks are her favourite, and if one makes an appearance on any of our journeys, no matter how far off they are or what our schedule might be, it's obligatory for me to take us off course in their direction. I motor to where they're sighted, keeping a respectful distance from the area, put the boat into neutral, and wait. That's my job. Hers is to watch them for as long as they're visible, oohing and ahhing over every fin flap. Since humpbacks' pectoral fins are the largest of all marine mammal appendages, and almost a third of the whale's body length, this can be pretty amazing. Those humongous fins also contribute to humpbacks being the most acrobatic of all whales, despite their huge size. When they breach, I admit, it's breathtaking.

I immediately did my part this time, carefully adjusting our course toward the whales about a mile or so in the distance. I felt excited too. Finally—a genuine wildlife encounter on this trip! I knew it would be a badly needed shot in the arm for poor Lyza, whose spirits were pretty dampened after all that had happened in the last few days.

As I saw her skipping back to me from the bow, I could see the injection already doing its magic when I looked at her face. It was radiant. She burst out with, "I'm not sure yet, but I think we're going to see our first bubble feeding!" From the way she said it, I knew this was big news.

"What's bubble feeding?" I asked.

The incredulous look she gave me made me aware of my appalling ignorance on the subject.

"It's only about the rarest wildlife encounter anybody could ever see," she said, leaving off the "stupid" at the end, but she might as well have said it.

I forgave her tone when she hurriedly went on in that enthusiastic way she has when she gets excited. She explained how a group of humpbacks team up to catch a school of fish in a unique way. One of them swims around the school,

Cast Off Your Bow Lines

blowing air bubbles from its spout to create a large bubble net to contain the whole school. It's large enough for the rest of the whale's cohorts, anywhere from three to a dozen or more, to swim below the net and drive straight up in unison inside the circle, scooping up almost all the fish in their open mouths. It sounded pretty dramatic as she described it.

"It's really rare adaptive behaviour," she went on, obviously well-versed on the subject. "And they can do it over and over again when the fishing's good, all of them keeping their same positions every time they surface. Some estimates say only about a hundred or so among the world's humpbacks have figured out how to do it. Most of those who bubble feed are here on the northern coast of British Columbia or in Alaska. If we get to see bubble net feeding, it will top my list of wildlife encounters for sure."

That was a pretty significant statement coming from this girl who had actually had a fair number of spectacular nature experiences up to that point. Since then, over the years, we have viewed grizzlies ripping into a humpback carcass in Glacier Bay, Alaska, black bears catching salmon from a rushing waterfall, a pod of killer whales overtaking *Porpoise* under sail and escorting us for miles, and porpoises in schools up to a hundred strong in the open ocean. We've swum with spinner dolphins in the wild off the coast of Hawaii, snorkeled with more than a dozen huge green sea turtles at a time, watching them up close and personal as they tore their breakfast of algae off the rocks with their beaks, seen groups of adorable sea otters (another of Lyza's favourites) holding hands in a circle off the outside coast of Vancouver Island, and we've even made the acquaintance of a massive fin whale on our way to Alaska. From less than twenty feet off our beam, his eye gave us the once over, and we had felt the connection with this sentient marine being. But I'm pretty sure if you were to ask her today, those bubble feeding whales that day would still be at the top of her list.

As Lyza spoke, the water broke two hundred yards off our port side, and three enormous whales burst out of the water, half their body lengths rising straight up in the air. Fish streamed out each side of their enormous maws before they slowly sank back below the surface and were gone. Only a huge circular wave radiated out from the spot where they'd been. Clouds of seagulls screamed, flapped, and dove for leftovers in the troubled waters.

Needless to say, our photographers covered the deck, crouching down with cameras clicking furiously. Every scrap of their camera equipment was used to record the event, including a huge bazooka-like lens positioned off the front hatch like a small cannon. I really wanted this positive moment to last and was

The Charter from Hell

gratified to see that the whales seemed quite happy to continue to feed in this spot. For two hours they kept at it while we lingered, just drifting in the channel. We turned off the engine to be silent observers of this clever and breathtaking harvest. During a lull in the bubble feeding action, two of the graceful denizens of the deep timed their full breaches at the same moment, and Lyza thought she'd died and gone to heaven. Lars and his telephoto lens just happened to be in the right place, aiming precisely at the spot to capture this magnificent photo opportunity, and there are spectacular pictures to prove it.

Mesmerized by this display, I chanced to look down over the side and saw bubbles starting to appear about thirty feet from our boat. *What if we're in the middle of this emerging bubble circle?* I thought with sudden alarm. *What if the herring are taking refuge under our keel?* In their lust for herring, would the whales fail to take notice that we were in the path of their intricate bubble dance? I could just picture *Porpoise* toppling over on the backs of those monsters in the middle of a feeding frenzy. Now wouldn't that put the icing on the cake of this seemingly ill-fated charter!

Almost as soon as I had these thoughts in quick succession, the arc of the bubbles headed away from us, and the circle started forming a little safer distance off our beam. We held our breath and trained our lenses on the centre of the forming circle. This surfacing would be the closest one yet by the looks of it. Suddenly, the three humpback hunters rose up like monoliths and, for their final bow, crashed back into the water. The wake of their fall rolled into *Porpoise's* beam, rocking her from side to side as though she were shaking herself in trembling excitement. Perhaps she was related to these creatures after all. Perhaps they breached so close to us on purpose to honour her presence.

Normally I'm impatient to keep a charter moving to the next destination, but on this occasion, I wouldn't have minded if we'd stayed to watch them all day. But even Lyza acknowledged that the last close-up display was their final curtain call and didn't offer any protest when I swung the bow toward Klemtu once again.

Motoring the last mile, Lyza leaned over to whisper in my ear: "I would almost endure this charter all over again for the privilege of seeing what we got to see today." Then she ducked down the companionway to begin preparing another lunchtime spread. I hadn't the heart to tell her what I'd heard on the radio that morning about the upcoming weather event. I was still working out in my own mind how to break it to her and the rest of the crew as she washed and I dried the last of the lunch dishes.

73

Cast Off Your Bow Lines

The crew had wandered off the dock as they waited for us to finish the clean-up operation when suddenly the cranky crewmate poked her head down the companionway. Then she declared in an accent made thicker by anxiety, "We have to leave right away. We can't stay here. I can't miss my flight, and the weather is going to get terrible—I heard it from the native guide we met the other day. A hurricane is coming!"

She was shouting now, and her voice was tinged with hysteria. Lyza looked at me with widened eyes, and if eyes could talk, they'd be saying, "Do you have any idea what she's going on about now?" I knew I had to nip this in the bud for all concerned, so I called a crew moot around the table to discuss the situation.

First, I tried to allay the fear that had been whipped up by our First Nations buddy, who we'd already discovered was prone to hyperbole. I carefully explained the Beaufort wind force scale and how the winds for each category of wind speeds are described as breeze, gales, storm-force, etc. Unfortunately, the descriptor for winds above sixty-four knots is called hurricane force. It was true that the next storm's winds would probably exceed that speed, so we'd experience hurricane force winds.

"We'll be fine as long as we're tied up in a safe place by the time those winds hit. The storm is due tomorrow night, so we have some time to make that happen. I have an idea of where to go and so that we'll get to where we need to be to ensure nobody misses their flight back home to Europe. Unfortunately, it's just not safe to head up to the end of the remote inlets we were hoping to explore, and I know that will be a disappointment to you all."

Our crewmates looked uneasily at each other. Although they had remained below the night before, they now had firsthand experience of the possible perils of anchoring at the end of those wind tunnels. I could tell Lars felt honour bound to ask if there weren't some protected areas we could go to and still salvage something of the charter's aims.

"There is protected anchorage up in some of these places, but you never know just how the wind reacting to the mountain valleys will come to bear on our position, and we're talking about big winds. I hate to say it, but I believe we need to make for Shearwater near Bella Bella, where you fly out."

"But that's just what I'm saying," said the mother, nodding emphatically with satisfaction. "Let's get going. What are we waiting for?"

Lord, give me patience, I prayed to myself. Out loud, I said, "I know you feel worried, but it's getting to be late afternoon. It will be dark soon, and the conditions aren't right to head out. We have a six- or seven-hour journey through

The Charter from Hell

some pretty dicey places ahead of us, and I will not be pressured into doing something that puts the crew and ship at risk." In my most captainly manner, I continued. "We'll stay in Klemtu tonight and leave early in the morning once it's light enough. That will get us to Shearwater in plenty of time to get secured. After the storm passes, we'll spend our last day exploring the area near there." And that was the end of that. They didn't salute, but I felt that my decision was respected by most of the crew.

Next morning, we departed after breakfast to make the run ahead of the storm. On the route, our difficult crew member chose to sit on the cockpit seat beside me as we threaded our way through a number of intricate channels. In a particularly tricky place, a narrow channel passed between exposed rocks as the building seas broke upon them. To the uninitiated, it would seem a very unnerving place. I was grateful for my more than fifty years of navigating these dangerous coastal waters, which have been called "the graveyard of the Pacific" with good reason. As she eyed the turbulent reefs passing by only thirty feet off the beam of the boat on both sides, her comment was, "Well ... I guess you know what you're doing out here, don't you?" It was a bit grudging, but it was her first and only compliment, so I took it.

That was about all the positive communication she could spare, but it sure was better than the volume of complaints she'd been piling up. The conditions were bad enough without an unhappy, fault-finding person in our midst. I had watched Lyza particularly as she plied the lady with tea, homemade cookies, anything to make her as satisfied and as comfortable as possible. I couldn't remember hearing even one thank you from her for anything the entire week, although the meals and the service were topnotch. I knew Lyza was at the end of her rope with this woman, and then there was another complaint about a drip finding its way onto the corner of her blanket. This was an annoying occurrence, Lyza conceded. We had done our best, but an occasional wind-driven drop would still occasionally penetrate the overloaded defenses. Then the lady started in again with her mantra of, "I've paid good money to be here and these kinds of things shouldn't happen."

That was it. Lyza put up her hand.

"Okay, this is what we'll do for you. They have very nice rooms in Shearwater, and we'll pay to put you and your daughter in one for the last two nights of the trip. How would that suit you?"

The woman looked genuinely taken aback at the suggestion and started stammering about how that wouldn't be necessary. But Lyza pressed her advantage. "Well, the offer stands if you change your mind, but if you decide to stay, I'd greatly appreciate it if you and your daughter would keep your complaints to yourself until the end of the charter. We're all trying to make the best of a difficult situation, and we need to work together to that end."

Lyza is normally the most gracious hostess you could ever hope to have, and she usually gets a lot of pleasure from serving our guests, but she'd been breathing mutinous threats to me in private for a few days before this incident. Thankfully, there wasn't another complaint the rest of the trip from mother and daughter after that little talking-to, and no more seditious mutterings from my wife either.

We rushed with the tide down Seaforth Channel and tucked into Shearwater before dark. Amazingly, considering all the boats that had taken safety in this fairly protected basin, we were able to secure a dock space. The seriousness of the situation was underscored when we saw that a tug had been brought in by the marina to provide a brace for the docks against the blow. In the previous storm, which hadn't been as bad as this one was predicted to be, the docks had become loose.

We heard the tug's engine chugging with its nose into the dock by midnight when the blast of the southeasterly struck. Gusts steadily increased to peak at seventy mph as recorded on our Windex indicator. It felt as if a great hand reached out of the sky to press *Porpoise* down, pinning her against the dock with every strong gust. I could only thank God we weren't out in some remote inlet. Knowing we were safe in port, Lyza and I relaxed and even went topside to experience the full impact of such an unusually powerful wind.

The tumult was over the next day. The surrounding water was as placid as if there had never been anything to worry about. It was our last day with our charter guests, so we poked around the environs of Lama Passage to find something of interest—a far cry from stalking the elusive spirit bear, I'm afraid. But we all made the best of it, especially the kind couple and the guide, who were very understanding and grateful for our care of them. Even the demanding duo seemed to moderate their tone to end the cruise on a positive note. Their (and our) deliverance was drawing nigh. It's true that uneventful, pleasant trips are soon forgotten, but the harrowing ones live in infamy, serving as colourful tales to tell our friends and relatives.

When our crew debarked the next day, we high fived each other, a ritual after every charter, and hightailed our way south toward home. On the way down we heard of yet another weather event that was bearing down on the coast within the next two days. We ran to beat the storm and finally, just as darkness was falling,

The Charter from Hell

slipped into the safety of the little marina in Port McNeill on the north end of Vancouver Island. As the fourth low pressure system in nine days whistled over our heads that night, I turned to my beloved as we cozily snuggled in the warmth of the salon.

"Lars said to me at the airport as he was leaving that he might like to try again next year. Would I be right in thinking you wouldn't be very open to that idea?"

A verbal response was unnecessary, as I read very clearly the deep antipathy in her blue eyes to even the thought of such a suggestion. There are limits to how far you should push a good woman. I was lucky to have a wife who loved sailing, who had hosted many a heavenly charter over the years, and who was the one to instigate the crossing of the Pacific Ocean. Discretion was definitely the better part of valour, and I had no intention of asking her to revisit hell. To tell the truth, I wasn't very anxious myself to sign up again for such an assignment. More than ever, I appreciated our sunny stomping grounds of the south coast of British Columbia.

Cast Off Your Bow Lines

PPSS

...We cannot make bargains for blisses,
Nor catch them like fishes in nets;
And sometimes the thing our life misses
Helps more than the thing which it gets...

                    Excerpt from "Nobility" by Alice Cary

This magic moment made it worth it all

8. The Call of Mother Ocean

Now we had done it. Now we were committed beyond the point of no return. We couldn't turn around and head back home at this point … not if I were to retain any pride at all. Not after months of preparation and expense. Not after pondering and planning this offshore trip, stepping into the big leagues of sailing adventure. Years of ever-increasing breadth of experience had led us to this anchorage, and now all that lay between us and the reality of actually doing the dream was a narrow little band of rocks—the breakwater protecting Neah Bay.

Neah Bay is a small, mostly native community on the northwest tip of the Olympic Peninsula, and for us it was the last outpost of civilization before the final plunge into the ocean. Early the next morning, we'd leave our safe harbour and sail into that great mythical Pacific Ocean I'd heard of all my life.

From my earliest years, I'd been aware of the big ocean presence that lay beyond my limited horizons. We used to camp on the Olympic Peninsula with my mother and her twin-like sister, Aunt Nuky. For whatever reason, our mothers, not the fathers, took my siblings and cousins on those fifties-era camping trips on the Washington or British Columbia coasts. My mother and aunt were so connected to the woods and wildlife of the Pacific Northwest that they were nicknamed Flora and Fauna. Regaling us children with constant chatter about the genus and species of a myriad of plant life, and decrying the depredations of loggers and land developers, they were the most passionate about exposing their children to the coastal waters and wilderness.

The trips in March and April to the wild ocean beaches of Lake Ozette particularly stuck in my memory. Just the smell of old canvas had the power to instantly transport me back to those damp campsites improvised among the dunes, where we braced with the bent evergreens against the incessant southwesterly gales. Our constant companion was the sonorous song of breakers thundering on the beach and crashing with a gravelly growl as they sucked the sand and pebbles into the undertow and soon cast them up again. The ancient sand dunes covered

Cast Off Your Bow Lines

with scruffy evergreens and coarse grasses bore witness to the eternal contention between land and sea. Whether we hiked the long, flat sandy beaches searching among the shoreline drift for Japanese glass floats or tramped on the mossy carpets in the misty rainforest, or snuggled deep in our army surplus mummy bags, that call of the deep resonated through the day and seeped into our dreams at night, sometimes a comforting presence and sometimes a booming terror.

I was always aware of the brooding ocean that lay on the other side of the mountains, but the natural barriers of Vancouver Island and the Olympic Peninsula kept it at a respectful distance. I knew her as the mother and progenitor of all those wondrous waterways running from the southern reaches of Puget Sound up through the American San Juans, the Canadian Gulf Islands, and on northwestward through the watery passageways to Alaska. These were the inland seas that safely defined a lifetime of cruising adventures and kept us from the massive ocean swells. Up until we owned *Porpoise*, we had only briefly been exposed to open ocean when we raced past Cape Caution on a calm day to cross the fifty-mile open stretch between Vancouver Island and Calvert Island on our way to reach the central British Columbia coast. That was our first introduction to the unfamiliar sensation of ocean billows that made our little double-ender sink out of sight between hills of water in Queen Charlotte Sound. It made an unforgettable impression.

Now anchored safely behind this last breakwater at Neah Bay under a serene sunset in early May, I was keenly aware as I watched the orange ball slip below the horizon that a firsthand encounter with Mother could be delayed no longer. Our avoidance of the big water was at an end, and the boundless deep awaited. Boundless! Even the word frightened me. No land in sight, no reference points to mark our position, no respite from those mountainous seas, no safe harbours to flee to for a calm anchorage at night. I couldn't help but think of past vacation views of unfettered ocean as seen from the windward side of Maui. I remembered how glad I was then not to be out there on any boat in that frightening, mesmerizing melee of tropical seas clashing with each other and smashing with explosions of white foam upon the jagged black lava. If anyone had told me back then that I'd be travelling in a sailboat 2600 miles across just such an ocean, I would have laughed out loud.

The prospect had a certain allure for sure, but the idea touched a cold nerve of fear. I was deeply conflicted, emotionally aware that the matriarch of the world's great waters could be a terror and a tyrant as well as a beneficent nurturer of all the rich and abundant water world that had been a delight all my

The Call of Mother Ocean

life. Whatever my feelings, this rendezvous was inevitable, and I would have to come to grips with this up close and personal encounter. Never had our charter business's name, Pacific Encounters, seemed more appropriate. It was time to step up to the plate.

As I pondered this impending meeting, I realized it wasn't just the prospect of the open seas that was fueling my angst. As the sun receded and twilight deepened, the shredded remnants of cloud turned from orange to magenta to grey, and I thought of how the horizons of our life on shore had taken on a darker hue over the preceding decade.

For almost thirty years, our life had been relatively clear sailing. There were a few storms and some delicate navigational challenges, but for the most part, we had an engaging occupation and a warm circle of family and friends who lent to us a tremendous sense of security, love, and purpose. We were confident of our bearings, knew where we were going, and had accomplished many positive things. Then our life experience rather rapidly changed and deteriorated. Our hearts were broken, our confidence was shaken, and we became keenly aware of our vulnerability in a world that had gone from safe and secure to ominous and chaotic. Our fundamental faith remained intact, but we were deeply rattled. We felt unsure of our compass heading, adrift in treacherous seas, over our head in deep waters. In a word, we found ourselves offshore.

This feeling wasn't unfamiliar. It wasn't our first time launching into the unsettling unknown. Many times in our life journey we had left safe harbours to step into uncharted frontiers. Lyza and I were only teenagers when we got married, left the support of family and friends, and moved away from our hometowns near Seattle to settle in Canada. We plunged with abandonment into the late sixties' radical Christianity that emerged out of the turbulent hippy era. At the same time and because of our belief in the guidance of a loving God, we cast off the normal means of making a living and chose to trust ourselves to divine providence, praying for our food, rent, and college tuition. We had no choice but to rely on prayer, as we were living in Canada without landed immigrant status and couldn't legally work for remuneration. Over a two-year period, we discovered it was possible to stay afloat and live by faith alone.

We opened a drop-in coffeehouse for the local youth without any backing from established sources. We were young and naïve but willing to step out and take risks. This pattern of unorthodox ventures continued to be our modus operandi in ensuing years. We joined a small interdenominational church and over time became pastors. Both the vocation and its location in Canada were a launch

Cast Off Your Bow Lines

into a totally new realm that was radically foreign to our previous life experience, and many were the interesting exploits of those years. Then after thirty-five years it all ended. We never saw this pastoral calling in life coming in the first place, and when it came to a close, we couldn't have anticipated it going.

We were in our late fifties, and just when it might seem it was time to slow down and start thinking about retirement, this new leap into Pacific Encounters came about. Five years later, here we were the night before casting off our bow lines to head across the great Pacific to the islands of Hawaii. Talk about launching out into the unknown! I had to acknowledge that in all our other ventures, although it didn't always seem that way at the time, we had been sustained and enriched. I had to trust that this voyage would have the same result. The journey began not in Neah Bay but back in the boat harbour of Blaine Marina, where our offshore journals began as we left the White Rock pier and headed out.

The Call of Mother Ocean

PPSS

… Behind him lay the gray Azores,
Behind the Gates of Hercules;
Before him not the ghost of shores,
Before him only shoreless seas.
The good mate said: "Now must we pray,
For lo! The very stars are gone.
Brave Admiral, speak, what shall I say?"
Why, say, 'Sail on! Sail on! And on!"

Excerpt from "Columbus"
By Joaquin Miller

Book Three

Catch the Trade
Winds in Your Sails

A Note to the Reader

After five years of strengthening our sailing and chartering skills up and down the British Columbia coast from the Gulf Islands to Bella Bella, we were ready to raise the bar of our Pacific encounters, and Lyza and I decided to sail *Porpoise* to the Hawaiian Islands. In May of 2010, our son, Matt, our son-in-law, Chris, and our new friend, Jack, joined us as we embarked on our first offshore voyage after a lifetime of coastal cruising.

We planned to take off from Juan de Fuca Strait and head south down the coast of Washington, Oregon, and California, sailing five to six hundred miles offshore. At about the latitude of San Francisco, we'd turn west and catch the trade winds to Hawaii. After exploring all the major islands of the Hawaiian chain for six weeks, we'd depart from Kauai to sail up toward Alaska over the North Pacific High, catching the prevailing westerlies back to Sitka, Alaska.

After spending most of August exploring the panhandle of southeast Alaska, we would make our way back down the Inside Passage to Vancouver and home to White Rock. The whole trip would take five months. We felt this was a voyage that could embody the spirit of our Mark Twain mantra. It would push us beyond our comfort zone, become an experience of a lifetime, and establish a milestone in the family sailing saga.

The trip was probably going to be a one-off venture, so we felt a journal would be an important way to memorialize it for future generations of sailors and armchair enthusiasts among our circle. To chronicle this adventure, I'd write an extended ship's log, a journal of sorts, sending entries back to family and friends from mid-ocean, allowing them to share vicariously in our escapades and get a sense of immediate discovery along with us.

For the uninitiated, I should mention a word on cruising logs. The ship's log has evolved into somewhat of an art form over the last sixty years of Clarke sailing trips. The practical purpose of a ship's log is to record engine hours, ports of call, wind speed observations, customs entry information, and all that sort of stuff. The earliest family boating logs were fairly boring, filled only with technical information, such as my Uncle George's observations aboard *Nor'wester*, painted grey during the war years, as he scanned the waters for Japanese submarines. I think my mother's influence on our family vacations in the fifties began to turn the *Nor's* log into a more entertaining enterprise. Her entries were embellished

A Note to the Readers

with artwork, humorous anecdotes, an account of the flora and fauna, and a record of the daily seafood harvest.

Her sons and daughter carried on this hallowed tradition when they started skippering the *Fred Free*. Each young family kept up the log in style, reading the entries of the previous crew, and trying to outdo each other in making it entertaining. These logs are all preserved and gathering dust on our shelves. They serve as books of remembrance on a rainy day and will be a way of keeping our brains alive by rereading them in doddering old age. These precious journals chronicle over eighty years of family sailing trips. The tradition continues on *Porpoise* as I conscript crew members, be they friends or family, to record their impressions and contribute their creativity. The current logs on *Porpoise* are replete with some pretty awesome artworks that vastly outshine my usual stick figure drawings. They also include some darn good poetry, many spiritual meditations, and, of course, bragging rights over any significant catch of fish. Included later in this book is an excerpt from my mother's log from the summer of 1952 when she cruised with her husband and four small boys. Our offshore journal is limited mostly to the open ocean part of this trip, as the offshore portion was all new to us and merited special attention.

I tried to edit out the more mundane log entry type of stuff, such as engine hours, weather observations, and latitude/longitude fixes, that often fill the pages of offshore voyagers. Our friends back home could follow our progress on our SPOT, a personal GPS tracker by which they could mark our daily latitude and longitude position on a chart of the Pacific Ocean.

Our ocean voyage ship's log focused more on our first-time impressions regarding this strange new environment and revealed some of our emotions and reactions to this mystic ocean world as seen from the deck of a recreational-sized sailing vessel. I tried to keep in mind that most of our friends and family would likely never do this kind of thing, and I wanted to help them to live in our shoes and go where only a few people go. Here are those same ocean journals just as they came to our friends back home. (Well, with a fair amount of editing since.)

Cast Off Your Bow Lines

Under sail most of the passage

9. Bon Voyage

Friday, May 7, 2010

Dear friends,

Good morning and goodbye. Yesterday we waved adieu to our kids and grandkids gathered on the good old White Rock pier and turned the bow of *Porpoise* southwest toward Victoria, the Olympic Peninsula, and the great Pacific Ocean beyond. We have sailed away from that pier many times, but never with the design we now have in mind to press past our familiar landmarks and commit ourselves to that endless horizon of the great deep.

I feel a little like the oxen that pulled the cart carrying the Ark of the Covenant back to Israel from the Philistine territories, lowing for their calves at home while their bodies kept moving along the path of their holy mission. The holy ark of our mission to sail the Pacific propels us westward, while another part of me clings to the familiar landmarks of coastal British Columbia. I am definitely conflicted between my keen desire to make this voyage and my deep-rooted apprehensions as I contemplate what we're about to do.

It's six o'clock in the morning. Lyza is asleep below, and Jack (whom I'll tell you about later) is at the wheel while I peck out a first journal entry to all you friends and armchair sailors who expressed an interest in following our voyage. I'll try to make regular entries in this journal, although I'm not sure how I'll get them to you. At this moment cell reception is still good as we pass Victoria heading out Juan de Fuca Strait, so I'm hurrying to get off a last word (I hope not literally) before the ocean swallows us.

Last night we made it to Mackaye Harbor on the southern tip of Lopez Island so that we could get an early jump on daylight and the tide. Jack, our guide, encouraged us to get up at one o'clock in the morning to get a good start on making it to our last anchorage of Neah Bay. He thought it would be great offshore preparation for the crew to get a taste of night travel and setting watches.

Cast Off Your Bow Lines

Of course, Jack planned to stay in the sack while the rest of the crew gained this valuable experience. I pondered his advice, took one look at the darkening sky, screwed up my face, and let Jack know we would defer the exercise and stay in bed, thus asserting for the first time my ultimate authority as captain. I turned in and set the alarm for a more reasonable 4:00 a.m. I figured night watches could come when they came. Besides, the night before our departure the rest of the crew, Lyza, Matt, Chris, and I had worked until two in the morning stuffing three tons of gear and groceries on board. We needed a good sleep.

Naturally, every crew member had overspent and overpacked, so it became a Herculean task to haul our gear and supplies aboard, toss them below, and then wade through the resulting piles of flotsam and jetsam while straining our brains to figure out where it could all be stowed. It's amazing the useless things you get talked into buying or bringing when ocean survival is at stake and in the back of your mind. On the one hand, you can't forget something crucial, such as sunglasses or a life raft, but space is at a premium, and you don't want to be tripping over some useless thing like a boogie board or an arctic snow suit. Nonetheless, no one was innocent of squirreling away some cherished personal effect or secret stash of their favourite indulgence. Hence, I looked the other way after catching the back view of Jack disappearing into his stateroom with three large bottles of Navy rum.

The Most Useless Item Brought Aboard award went to Matt, whose old boss at the Mountain Magic store conned him into spending twenty-seven dollars on a giant jar of powdered protein. On second thought, maybe the award should have gone to Jack for bringing several large packets of dried egg protein from his Vanderpol egg business. These items could be justified in the interest of survival on a life raft but were hardly useful otherwise. Somehow, the prospect of cramming dry egg or protein powder into one's parched mouth while languishing on a life raft under a scorching tropical sun wasn't very appealing.

On the other hand, having a life raft onboard is an absolute necessity, making it one of the most costly items on the essential-but-most-likely-to-never-use list. That list gets more extensive and expensive the more you think about all the eventualities in a worst-case scenario, so it's not good to read too many disaster and survival at sea stories—of which there are plenty. No one wants to hear about being lost at sea on a life raft for seventy-two days, fishing with twine and bent nails for evasive fish while having to drink your own urine. You know—"Water, water everywhere, nor any drop to drink."

90

Bon Voyage

With stories like these lingering in the back of your mind, a lot of money drains from the cruising kitty toward the better-safe-than-sorry category of the budget. Being a financial conservative, I found the expense side of the ledger more and more unsettling. After shelling out more than a thousand dollars on a used life raft, eight hundred dollars on an EPIRB (emergency position-indicating radio beacon), and over five hundred dollars on medical supplies fit for a military encounter, I finally put my foot down when it came to six thousand dollars to Lloyd's of London for offshore boat insurance. We decided to leave Lloyd's to insure tankers and the like while we risked it all. I just couldn't abide the thought of paying good money for that unlikely event you want to avoid at all costs, especially when a sinking at sea would likely be a fatal event. We figured it would be better to beef up our life insurance rather than our boat insurance, so at least our kids could benefit from our demise. Rather than replacing the boat, they'd get the settlement and take a vacation to Hawaii, paying respects as they looked down from thirty thousand feet upon those endless billowing waves. Even without paying extra boat insurance premiums, it cost thousands of dollars just to shore up our defences against all the eventualities we could think of. It's a reality check to consider that it all depends on your own pre-emptive preparations to make sure every piece of equipment is in top shape and that the means to repair or replace it is on board. Self-reliance is taken to a whole new level when surrounded only by thousands of miles of ocean on every side.

So here we are, loaded with gear and groceries and making good time on the ebb tide, which empties all of Puget Sound and Georgia Strait through this amazing Juan be Fuca waterway into the Pacific. We're sliding along with the snowy Olympic peninsula to the south and beautiful Victoria to the north, sweeping along at a good clip toward that ultimate destination—The Ocean. Needless to say, we're pretty jazzed. What will it be like sailing on the mighty main? The weather window looks fine according to our last check. No major depressions on the horizon—British Columbia woke to the first completely blue sky that I can remember this spring.

There was a howling cold wind with incessant rain showers every day in the Blaine marina while we worked at the final chores of ship preparation. Visualize Matt, if you will, suspended forty feet aloft in a bosun chair. That's marine jargon for a canvas-covered seat, rigged to hold you secure at the end of a halyard (rope that goes up a mast) while someone hoists you up amidst the shrouds and spreaders in order to do some necessary work. As you can imagine, this is no easy feat out in the great ocean rollers, where one has to become a sort of trapeze

Cast Off Your Bow Lines

artist to swing among the shrouds, climb over the spreaders, and get to the top and back down without bashing your body against the mast or flying crazily out over the water. Much better to get those chores done while still tied to the dock. Even there, Matt had to install a tri-light on top of the mizzen mast with glue and twine in the thirty knot gusts of a freezing north wind. The cold, hostile weather of spring did nothing to give us a sense of assurance about our coming voyage.

Today, however, this blue horizon looks much more promising, and the sparkling sunshine on the water is a tonic to our spirits. The wind in Juan de Fuca is usually westerly and often bellicose. Perhaps this is its way of warning us to stay safely inside Vancouver Island. On this fine day in May, the wind is strangely filling our sails with a gentle breeze from the east, as though ushering us kindly toward the Pacific. Jack is befuddled with such luck, letting us know that he has never, in all his many travels, done anything but bash his way out of the Strait of Juan de Fuca with the wind in his teeth.

"It's a great sail coming back but a challenge to get out, especially when you hit the Swiftsure banks where the tide meets the ocean swells." Jack's trademark grimace transforms into a sunny, good-natured grin as he says this, He then proceeds to prophesy: "But this is a lucky ship … a lucky ship," he laughs.

"How do you know that?" I ask.

He just laughs again. "Oh I know …I know…"

We can see sports fishermen in the distance bobbing off Race Rocks, probably mooching for salmon, and two or three freighters in a line are plying their trade into the northwest ports of Seattle or Vancouver. I can see Cape Flattery in the distance at the end of a snowy march of Olympic peaks. It looks uncharacteristically benign today, and I'm aware that our last quiet repose before our leap into the void lies at the base of that headland.

So bon voyage! No, wait a minute, that's what you say to us. How about sayonara? Maybe not—that carries a certain note of finality with it. I hope not! I hope we haven't forgotten some vital thing. I'm a little worried that we haven't stored enough coffee. Some things are absolute essentials, you know, and there are no Starbucks where we're going. There's no nothing. It all has to be somewhere on this little life boat—our world, our space capsule for the next three or four weeks before we see land again.

Signing off … Captain Clarke (not to be confused with Cook!)

Bon Voyage

PPSS

"There, sooner or later, the ships of all seafaring nations arrive; and there, at its destined hour, the ship of my choice will let go its anchor. I shall take my time, I shall tarry and bide, till at last the right one lies waiting for me, warped out into midstream, loaded low, her bowsprit pointing down. I shall slip on board, by boat or along hawser; and then one morning I shall wake to the song and tramp of the sailors, the clink of the capstan, and the rattle of the anchor-chain coming merrily in. We shall break out the jib and the foresail, the white houses on the side will glide slowly past us as she gathers steering-way, and the voyage will have begun! As she forges towards the headland she will clothe herself with canvas; and then, once outside, the sounding slap of great green seas as she heels to the wind, pointing South!"

Sea Rat in Wind in the Willows
By Kenneth Grahame

Preparations for going to sea

10. Juan de Puca

May 9

Okay, so here I am before the mast, so to speak, literally straddling the fore-hatch with my feet while balancing my butt on the deck with two fenders stuffed under either thigh. My Macbook is clenched between my knees as I try to write a brief report, and seven or eight foot billows are rolling under the bow as we pitch back and forth. Attempting to fulfill this task is like trying to write on a roller coaster. I wonder if it's worth the effort to give you a firsthand account, especially when the sea is so rambunctious, but I guess if a war correspondent can report in the midst of battle, I can do this.

It's day two of open ocean sailing. Thought you all might want to know about our initial reactions and the state of the crew. I can report that one of us is laid out on the cockpit bench under a blankey. He has hardly moved from this spot since early yesterday, when we met the first ocean swells. They were rolling up against the tide onto Swiftsure Bank, a shallow area off the mouth of the Strait of Juan de Fuca, or maybe we should now call it Juan de Puca, in honour of poor Christopher. Last night after contributing to the strait once again (fifth time … I was counting), he collapsed on the stern bench and melded into the cushions, exhaling a weak groan. In humble submission to his fate, he remained there all the long, cold night without so much as a whimper. We found him in the morning looking, in *Princess Bride* terms, "not all dead but mostly dead." He was pale and trussed up like an Egyptian mummy in heavy weather gear and blankets. This might have been all right if he could have taken the night watch at the same time, but, alas, he could not. He was indeed a useless mummy.

Night watches—those three-hour stints enclosed in the dodger cubicle, rocking in the helmsman chair, staring at the red numbers of the compass. Hey, it's dark out here at night—very dark outside this tiny cubicle. When you stick your head out from behind the flimsy protection of the dodger, there's only the

Juan de Puca

wind blowing on your face and the sound of the waves splashing against the hull. Under full sail without a motor running, there are no other sounds, and *Porpoise* surges into the black void, heedless of what lies ahead. We can only hope there are no drift logs of the kind we encounter on the coast, or, worse, a loose, partly submerged shipping container acting like a portable reef—a danger our forbears did not need to heed.

The freighters running in and out of the coastal ports are the more imminent dangers. Right now we're still in the Seattle/Vancouver shipping lanes, and it's up to us to keep an eye out for big ships by watching for their lights in the dark or by detecting them on our GPS screen. Thankfully, we're equipped with AIS technology (another pricy but necessary expense) that alerts us to other ships that show as red triangles on the screen. *Porpoise* will also appear on those ships' screens as a red triangle to warn them of our insignificant presence, assuming that one of their crew is actually on watch and not distracted or playing poker down below. It behooves us to be watchful, as there is no question who would fare the worst in a collision. They might not even hear the smash-bumpety-crunch of our mangled remains sliding under their keel. On the brighter side, since there is more than a mile of water under our bottom, at least we don't have to keep watch for rocks, reefs, and headlands out here like we do when coastal sailing.

One after another we hunker down alone on the helm seat, watching over our loved ones as they sleep below on their narrow bunks in the belly of our ship. We focus on the little electronic images of the compass and the chart plotter, trusting by faith that they tell us the truth. There's also the glowing green "16" on the VHF radio. Nearer to land, the coast guard monitors this channel for mariners in distress, but we're already too far out for that. Occasionally a voice comes out of the box like an oracle, startlingly loud in the dark, but it will be from a freighter or occasional fishing boat somewhere in a thirty-mile radius from our location. It's kind of spooky when it happens but also comforting to hear another voice and know that we're not the only ones dumb enough to be out here. However, it's mostly commercial traffic this far out, and most of them aren't very chatty, instead speaking in impatient, clipped phrases if you say anything beyond technical information. We hope to meet someone more friendly before we're done.

This is a strange business for us. We veteran inland water cruisers don't take readily to this ascetic business of staying awake at all hours while squinting through tired, drooping eyelids trying to locate freighter lights. A coastal sailor is used to the knowledge that, after a hard day of sailing and fighting it out with

Cast Off Your Bow Lines

mother nature, you get to find a secure cove to anchor in, sit back with a cuppa or a pint, and leisurely watch the sunset. You know what I mean? There's supposed to be a definite end of the journey, a satisfying consummation of the day's labours, a restful time of meditation alone or quiet convivial conversation with others to bring closure to the day. But out here the sun goes down, the rollers keep rolling, the wind keeps blowing, the sails are all still up, and so are we. It feels like there's something wrong when we don't stop at all and we realize that this will be the story for us tomorrow, the next day, and for days and days. This isn't just a little visit with Mother Ocean. We've moved in and are living with her. "Are you having fun yet?" you ask. Well … yeah … I think so … but there are some serious mental and emotional adjustments involved.

Perhaps I'm making this sound too grim by giving you the scary adjustment side of things. Actually, our first night on the ocean was also quite magical. Matt took the first watch from eight to midnight. Well, he pretty well took the whole first night, as Captain Cook … er, Clarke, the adventurer, salty seaman extraordinaire, was feeling a little … well … just a little queasy. Matt's brother-in-law, Chris, was down for the count though sometimes awake on the cockpit bench. Few crew members could claim total exemption from the curse of Juan.

I was sleeping it off after dinner for a few hours on the first night when I heard Matt calling me to come up on deck. Hoisting myself upright from the bench below, where I had collapsed, I shook my head to come out of my chaotic dreams into the chaotic reality of *Porpoise* rolling through the swells in the pitch black. I looked at the clock. It was 2:00 a.m. Matt was peering down the companionway and speaking to me out of the night.

"Dad! There's a freighter crossing our track out here. I think you better come check it out."

Groggily, I followed Matt out on deck. The cold wind slapped me in the face as I stumbled my way to the cockpit, where I clutched the stern rail and thumped down on the seat. Matt greeted me genially (he doesn't experience seasickness), and Chris grunted from under his blanket. Glancing about, I took in the magic of my first night at sea. Staring astern, my attention was irresistibly drawn to our phosphorescent track. Glowing turbulence swirled up in a corkscrew of bright sparks from underneath the keel, the trail left by a ship under sail, silent and ghostly. I squinted at the constellation of organisms in the sea and then gazed up at their mirror image in the sky. The Milky Way marked a pathway of white across the heavens in such clarity that it appeared a trail of glittering white coals. It seemed as though we floated not on a boat in the water but suspended

Juan de Puca

in space, with the stars surrounding us above and below. Small clouds, moving past us with surprising speed in the brisk night wind, appeared as dark spots that momentarily blotted out patches of stars.

My initial recoil from the bracing wind vanished, giving way to numb enchantment, the same sensation one feels after an invigorating plunge in cold water—the first shock and then a refreshing delight. This introduction to the night realm of the offshore mariner thrilled me to the core. To sail a ship through the night, at least when the weather is tolerably fair, is magic. The motor is silenced, the only sounds are the wind in your ears, the flap slap of the sails making time, and the wild, galloping cadence of the vessel plunging its way through the seas. It was as though we were riding the back of a living creature in some mythological tale. I suppose it's old hat for seasoned sailors, but for us it is a wonder, and I doubt it will soon lose its chill and thrill.

We sat absorbing it all until Matt informed me there were signs of visitors trailing behind us in our wake.

"I can't tell what they are. It's too dark, but we've been hearing something for the last while, maybe dolphins."

Straining my eyes to penetrate the darkness, I saw nothing at first, but as my eyes grew used to the dark, I could discern ghostly trails of bioluminescence zigging and zagging through the sea. Then I heard the sharp inhale and exhale of mammal breath huffing and chuffing in the dark. There had to be at least two, maybe three, large creatures trailing us. Dolphins? Porpoises? Maybe they were making the acquaintance of the mother ship named in their honour. Whatever they were, they didn't seem to be in a hurry to go anywhere and kept us company for some time. Were they being friendly or just using the lift from our keel to ease their passage? I prefer the more personable answer. I say they do it to be sociable, which is confirmed by that perpetual smile fixed on their faces when seen in the daylight.

Matt reluctantly broke the spell of this spectacle behind us to draw our attention to what lay ahead of the bow. There it was on the screen, and soon we could hear the rumble and see the lights of a freighter ahead, but still a way off. In the context of our magic moment with nature off the stern, the apparition off our bow seemed like a stark intrusion of machinery and manmade light. The deep thrumming sound of its great engine could be heard over the water even at this distance. Although it was alarming to see it charging across our heading, we determined we were not on a collision course. It would definitely pass ahead of us. Whew! Still, it was an opportunity to try and make contact on the VHF radio, which brings out the boy in the heart of most men.

97

Cast Off Your Bow Lines

I felt an irresistible urge to pick up the hand receiver and say something cool like, "Calling all cars... calling all cars." Or maybe, "Roger that. Ten four, good buddy." I resisted the urge and instead said, "Calling the freighter ahead ... this is Sailing Vessel *Porpoise*. Do you read? Over."

After some static and an awkward silence, we finally heard a man's voice come through. He seemed interested enough in conversing with such peons as us. The captain assured us he knew we were there, confirming that our AIS was working. He told us he'd pass three miles ahead of us. We felt pretty grown up and pleased with ourselves to be in the big leagues now, in radio contact with the ocean freighter fraternity. He even knew the name of our vessel from reading it off his screen. We're on a learning curve out here, but it's exciting to get the hang of things.

After the baptism of that first dark night, it was a welcome sight to see the sky begin to brighten and then feel the sun's warmth, although it wasn't visible to us under the marine cloud cover. It feels fine to find ourselves charging forward toward Hawaii on these fresh seas, even though my legs and body are already tired of bracing against the constant movement. My head is beginning to buzz, and my stomach is feeling a bit queasy. I'm confident we'll get used to this.

Juan de Puca

PPSS

"My men grow mutinous day by day;
My men grow ghastly wan and weak."
The stout mate thought of home; a spray
Of salt wave washed his swarthy cheek.
"What shall I say, brave Admiral, say,
If we sight naught but seas at dawn?"
"Why, you shall say at break of day,
'Sail on! Sail on! Sail on! And on!'"

Excerpt from "Columbus"
By Joaquin Miller

Chris' introduction to Juan de Puca

11. Boggled

May 11

Okay, I'm back—two days later. You wouldn't believe how long it takes to do stuff out here. Every normal thing takes two or three times longer. Even now, here I am trying to write (in the same place as before … leaning against the main mast) with two medium-sized fenders crammed under my right and left thighs. Why? you say. Well, because the boat can lurch twenty to thirty degrees from side to side, so the fenders keep me from doing a barrel roll right off the deck to port or starboard, which would bring our great adventure to an abrupt and inglorious end. Also, the fenders help me sit cross-legged, yoga style, without discomfort to my stiff old legs. Not very ascetic, I know, but it makes me feel spiritual and allows me to imagine myself like Bernard Moitessier, the great French solo sailor. I always admired his famous posture, sitting cross-legged on top of his pilot house, sailing over the oceans, his long dark hair streaming back—a yogic flyer of the seven seas.

Even with the fender support, I have to constantly strain right and left to keep my balance. It's exhausting. Jack says a man can lose two inches off his waist after a few weeks out here because of the workout your core muscles get straining against the motion of the ship. I doubt there would be many takers for such a novel approach to weight loss. Obviously, these aren't ideal writing conditions. Hitting wrong letters as I type happens twice as often as usual. It's also hard to keep track of my mouse under this bright grey Pacific sky. After many minutes of searching, I finally located the little varmint hiding among the flowers on my desktop photo. I had to coax him out into the sunlight and gingerly guide him home to my designated application, knowing he'd disappear again shortly. I hope you appreciate what this is costing me to report to you.

So, you ask, "How is life at sea?" Well … it's great! Amazing, actually. It's such a totally different environment. It's a front row seat on glorious chaos. The

Boggled

sea has turned a deep indigo blue in the last two days. It looks as if it's been dyed this unnatural hue. Draw a pail of this water in our white bailing bucket and it actually looks like the water is tinted blue. I'm not kidding. The ocean is such an enchanting blue that as I sit here mesmerized on the bow, climbing with the ship to the peaks and crashing in the troughs, I almost feel I could cast myself overboard into Mother Ocean's billowing bosom … that is, until I get a cold splash of reality in the face from a slaphappy crest, which dashes this seductive, suicidal spell from my head. The open sea is an absolute wonder, especially when the sun shines and sparkles on the waters, and the white clouds are flying, and the wind's like a whetted knife …

Yesterday the sun shone and the winds blew twenty-five to thirty-five knots out of the northwest, and we rushed along under full sail and then reefed the main (that's sailor talk for reducing the amount of sail you have up). Then we took all the sails down except the genny (that's not a girl—it's the nickname of our large genoa sail) and flew along at top speed, twenty-six tons hauled through the water by this slight piece of cloth.

The seas built under this steady fresh northerly until they rose up as blue hills behind and before us. We sled our way down their faces, only to find them rising up before us again and leaving us to wonder if we were catching up to the seas ahead or if they rolled up from behind to overtake us. I'm pretty sure they come from behind, as they appeared to move forward to rollick and frolic with an infinity of others crashing and plunging in a great herd toward the fathomless horizon. It was as though we were galloping on horseback, picking our way through endless blue dunes that are always moving and dodging to toss us about and occasionally slap us in the face with a splash across the bow.

Periodically, a lonely shearwater comes over the horizon, soaring only a foot or so off the water, dodging and diving, expertly weaving his way while never touching the water. We think he searches for the little flying fish that have begun to appear. They are the little sea mice of our blue meadows who suddenly appear in tens and twenties, as though flushed from hiding to flee *Porpoise*, which to them must seem a mighty whale in hungry pursuit. The shearwater seems nonchalant in his flight and uncaring of our presence, but we know different. He stays with us a long time, racing up from behind to glide alongside and then crossing ahead of the bow to weave back and forth ahead of our track. Maybe his isolated existence is cheered by *Porpoise*, which may appear to the shearwater to be a giant seabird spreading her great white wings over the water in an attempt to become airborne. Maybe that's why he lingers, to see if we'll ever launch off

101

Cast Off Your Bow Lines

the great waves and into the sky to join him. Maybe he thinks the big bird is too weighted down with the odd creatures riding upon its back. Eventually convinced there will be no such ascent into the air, he moves off on his solitary way and we on ours. Still, he must be keeping tabs on us, for he renews the acquaintance the next day and the next.

I must admit to a growing love of this great ocean, although at the same time, I'm annoyed and aggravated by it. Like my grandson Kai (his name means "ocean" in four languages), it's never still and produces similar emotions. The ocean's convolutions are perpetual. Where I come from, the seas follow patterns of orderliness according to the prevailing wind, moderated and tamed by the coastal shorelines. The waves line up in straight lines to advance from north, south, east, or west so that a sailor can choose his point of sail and approach the wave train at a specific angle of comfort.

Out here it's a free-for-all. The constant swells come from far away, generated by distant gales and storms. Sometimes we get caught between wave trains from different directions that meet in a choppy melee. Over top of it all, the prevailing wind generates smaller waves among the swells. There is many a rude slam and bump as we pass through such a stretch and water slops over the deck and casts us about like sacks of potatoes. Eventually the confused seas sort themselves out into a more regular pattern. Maybe the ocean isn't always so raucous (I'm hoping not), but according to these first introductory days of life here, I have concluded that it's pretty much anarchy most of the time.

While we're all somewhat overwhelmed getting used to this new life at sea, Jack keeps us fearing for worse in the future by saying things like, "Ahh … well … yeah … this isn't much … just a nice wind off the quarter to keep us in trim. Wait until we have to go to windward." He delivers this with his trademark grimace followed by a cheery smile and chuckle to himself. "I've seen a lot worse, a lot worse."

The first time Jack sailed to Hawaii, he landed in a storm with hurricane force winds of at least sixty-five miles per hour, which forced him to remain hove to (parked, as he puts it) for three days.

"Wind like that presses your ship down like a great hand and hits you hard in the face like a sheet of plywood. It's like running into a solid wall."

So in just about any conditions we might face, Jack can always be heard to say, "I've seen worse … much worse." I definitely don't want to see any worse, thank you very much.

Boggled

This spirit of rowdy chaos offshore permeates every facet of our existence in this tiny floating community. Everything becomes a major challenge, including meal preparation, eating (a major preoccupation out here), sleeping, reading, going to the bathroom, and just generally travelling from one part of the ship to another. All these tasks require forethought, concentration, quick reflexes, and the balancing abilities of a gymnast.

Let me try to explain this extraordinary living experience to you more specifically so you really get what I mean. Yesterday I was elected to make breakfast. I decided to cook my patented blueberry pancakes with bacon and eggs. (That's right, we don't skimp out here. Just because life is insane for any normal stomach doesn't mean we're going to nibble at oatmeal gruel and hardtack every day. Just because it might all come back up doesn't mean it shouldn't go down with all the savour possible.) I can usually whip up pancakes from scratch in a jiffy, and I take great pleasure in the process—nothing like the smell of bacon sizzling and the sight of fluffy golden pancakes rising on the griddle.

The first task was to find, measure, and dump the dry ingredients into a stainless bowl precariously balanced on the narrow counter. The recipe is fixed in my head—two, two, two, two and two, one and one. That's two cups flour, two of milk, two eggs, two tablespoons of sugar, two of cooking oil, two teaspoons baking powder, one teaspoon of soda, and one of salt. Easy—when you don't have to dump and hold simultaneously. A maxim on the high seas states you must always leave one hand free to hold on to the boat. Not having a third hand left me in jeopardy.

One has only to glance above the counter on which the pancake batter sits to stare directly into the frothy seas rolling up toward the large pilothouse windows. These portals offer an intimate front row seat to the surrounding environment. This was an attractive feature in *Porpoise*'s design. The downside of this panoramic view when offshore is the constant and unnerving sight of huge waves barging right up to you. The biggest of these can be seen coming from a hundred yards away, and you watch them grow and grow as they roll up to the window. Suddenly, the wave is in your gawking face, a freight train of water that you expect to smash right through the little pane of plexiglass—the only thing that separates you from the threatening outside world. But to your surprise, the wave slides somewhere underneath with a lifting force that heaves everything to port, including you, forcing your head back and your eyes up toward the heavens, as if in supplication, before rolling you abruptly back into the ensuing trough, where for a terrifying moment you contemplate the abyss.

103

Cast Off Your Bow Lines

If you're the cook standing at the counter, the first motion presses your back against the four-inch-wide restraining belt fixed across the galley to prevent a body flying across the boat, and the second movement casts you back toward the windows, where you teeter on tiptoe over your bowls for a moment with the marked perception that there is nothing to prevent you from falling into the watery chasm ... or at least into the pancake batter.

You can imagine the watchfulness required to keep on top of the dishware on the table and the cooking stuff on the counter and stove. All comers for breakfast have pre-assigned seats and defined areas of surveillance. The tricky thing is to negotiate that seventh sneaky bigger wave after the six milder ones have made us complacent. When the big one arrives without warning, all hands grab at a bowl, plate, or steaming pan. When the full complement of eating and cooking ware exceeds the number of restraining hands, notable misfortunes occur. It's essential that every loose object be given a secure home to live in and be returned to, but this almost never happens because there are five un-trainable people on board (make that four, because the fifth one is Lyza) who must cooperate with this "place for everything and everything in its place" rule but never do.

So with that bit of context, I return to breakfast. Once the batter was finally mixed and the blueberries added, I fired the burners to begin cooking. I felt like a tree faller in a wind storm held in place by the restraining belt slung across my backside. In front of me sizzled the scalding hot bacon grease in a frying pan. Next to the bacon was the smoking pancake griddle, and next to that the batter bowl. Realizing I wanted a plate for the bacon, I made a move to push open the cupboard slider where a dozen or so Corning Ware plates were stacked. Suddenly, the ship lurched, and my left hand instinctively reached for the ceiling to grab a handhold while my right steadied the dangerous pan of bacon. Both hands occupied, I could only gape at the batter bowl as it slid and banged into the lip of the counter, where it momentarily teetered until the ship pitched violently back to send it slamming against the cupboards, slopping a wild smear of batter over the teak. As the ship pitched back again to port for the following wave, I decided to abandon the bacon to save the batter, which was sure to leap the parapet on the next run. As I lunged for the bowl and thrust my hip against the stove to pin the bacon, a white blur shuffled like a deck of cards directly under my nose. The following explosion of the Corning Ware in a tinkling crash down the stairway brought the immediate realization of my folly. I had left the cupboard slider open. Five or six plates were in a shambles on the floor. My right hand momentarily forsook the batter to fumble with the slider and slam it shut before any

Boggled

more plates could frisbee into space. Corning Ware is pretty tough stuff, but if it falls hard it doesn't just break—it shatters into a zillion pieces. The fragments covered the floor and the stairs and were scattered over the salon bedding. It took days for us to discover every errant shard, sometimes with our bare feet.

Feeling bested by the inexorable physics of the situation, I enlisted Matthew's help. We started counting seconds between the big wave trains. Okay ... wait ... wait ...wait ... now! Ladle the batter, flip the pancakes, grab a plate, hang over the belt, and pass it to waiting hands just before the boggle cubes are upended again. As we mastered this great ballet of breakfast creation and consumption, we managed to feed everyone. Then we'd begin the same careful dance for the challenge of clean-up. With double or triple the normal energy expended for such a simple matter as the morning meal, you can imagine why generally all we can do is service our basic needs and nap a lot.

Should I describe to you the bathroom challenges? Perhaps not in great detail, hmm? Suffice it to say that even macho men sit. Yes, sit. I know, to me the very idea seemed sick and wrong too. Surely with a little concentration it couldn't be too hard to hit the target, a reasonable enough art to master while at sea. However, it soon became apparent there were no such masters aboard, and when the complaints of the toilet cleaner reached a certain pitch, we surrendered to common sense and feminine sensibilities. Once the new potty procedure was put into practice, it was discovered to be a very beneficial habit in many respects, not least of which was the benefit of the restricted quarters of the head, which became one of the best places to be alone and find secure moorings against the constant motion.

All this talk brings up the subject you women would most be curious about. How is poor Lyza doing out there cooped up with four festy men and the conditions herein described? I am happy to inform you that Lyza has been awesome. She is thriving, even perky most of the time. She has an iron sea stomach, a cheerful disposition, and a servant's heart—all invaluable assets during a passage at sea. She brings the woman's touch of civility, order, and cleanliness into a situation that could easily degenerate into male squalor. Lyza is at the top of her game on the ocean in every way.

She quickly ascended in all our estimation as the most valuable crewmember and hardiest seadog of us all. By a unanimous vote, the men exempted her from night watch duty—not primarily out of chivalrous deference to her female gender, but more selfishly out of regard for our stomachs. It is much easier for us to face long, cold vigils at night than our own cooking, although some of us

105

Cast Off Your Bow Lines

make some pretty palatable contributions to the table. She, on the other hand, is indefatigable in her heroic stints at the galley stove, often three times a day, as well as doing her regular watches on the helm during daylight. She is the undisputed mainstay of ship morale through her enthusiasm for the adventure and her dedication to the ship's cuisine.

Even she, however, faced a few dark moments ... like yesterday when she forgot to gimble the stove, which is a gismo that enables the range to swing with the waves. I was stretched out on the pilothouse bench at the time, taking a quick queasy break from the helm, when a big wave hit. With a screech from Lyza, the half-cooked dinner, saucepan, and all flipped off the stove. It flew across the boat and exploded upside down, spraying hot chicken and curry sauce all over the floor, the bench, and me. Luckily, I was swathed in heavy weather gear, so I the scalding sauce didn't burn me. The next wave sent it all hurtling back in Lyza's direction, who, with great aplomb, fetched and forked the chicken off the floor and back into the pot, contrary to her usual Howard-Hughes-like aversion to germs. One has to lower the bar out here. Besides, a boatload of hungry people won't ask too many questions as long as the food is hot and tasty, which it was.

These are some of the challenges of life on the inside of the boat, but most of our time is spent outside on deck. This is a very outdoor experience, a fact made clear every time you step up on deck under the canopy of limitless sky and see miles and miles of ocean bounded only by the flat line of horizon on every side. Most of the time you want to be out on deck rather than below where you can't see the source of all the commotion. Down below, fear can become amplified by one's imaginations, whereas looking those seas straight in the eye reveals what you are up against and enables you to brace against the larger waves.

It's also better to be on deck in the battle with seasickness. The green person (whether by experience or facial hue) is put in the helm seat, where they can focus their eyes on the horizon, breathe in the fresh air, and feel as if they're taking control in a situation that feels out of control.

Chris spent many hours this way in his first struggles with mal de mer, clutching the wheel and staring grimly ahead. Now that it's trending a little warmer every day and we've gotten over the initial shock of this different existence, most of us have become quite healthy and even a little confident about finding our way around.

Up on deck there's more imminent danger of falling overboard, so precautions are made to keep everyone safe. One such safety measure involves the rigging of jack lines that are strung along the centreline of the ship. These have

Boggled

nothing to do with Jack; it's just what they're called. When you go forward, you clip on to the jack line with a tether and harness so that if you're pitched off the boat, you're held to the mother ship instead of quickly disappearing astern. There you can dangle by your tether and bang off the side of the boat until somebody notices you. They'll have to figure out how to haul your mangled remains back on board to revive you, or, failing that, they'll cut the strap and give you a proper burial at sea. Jack says we don't have to worry much about trying to rescue the man overboard because if anyone falls off, they're just dead, that's all, and the best thing is to sink quickly. No way will the crew be able to turn around a boat under full sail to come back and rescue you.

"Nobody will find you once you fall in out here. Better to sink fast and not prolong the agony."

We actually proved him wrong by practising a man overboard manoeuvre in mid-ocean and successfully rescuing a bucket. He was quite impressed, but it didn't change his views about sinking quickly. This is the same guy who told us not to worry about having a life raft. He figured it was just another way to pro-long the agony of dying at sea. Lyza had put a stop to that kind of talk back on the dock and made it quite clear she wasn't going offshore without all the proper safety gear, including a four-person inflatable life raft.

After hearing the Dutchman's survival philosophy, we all decided he could be crew member number five.

"Bye, Jack. While you sink, we'll try our luck with the raft." We suspected his wife, Judy, wouldn't subscribe to such negative nonsense and would have some harsh words for us if we came back without him.

"You just let him sink? Who's going to pay the mortgage now?"

Well, I need to sign off. There's so much more to tell you, but I'm sure it will eventually come out as we ramble along. I'll start to get seasick again if I stare at this screen too long. All the crew says hi.

Cast Off Your Bow Lines

PPSS

"Some went out on the sea in ships ...
They saw the works of the Lord,
his wonderful deeds in the deep.
For he spoke and stirred up a tempest
that lifted high the waves.
They mounted up to the heavens and
went down to the depths;
in their peril their courage melted away.
They reeled and staggered like drunkards;
they were at their wits' end.
Then they cried out to the Lord in their trouble,
and he brought them out of their distress."

—Ps 107:23–28

Six hundred miles offshore

12. There Be Treasure

May 14

Three days more and three days wiser. I fear my last entry about being boggled may have left a more negative impression than is fair to Mother Ocean. My initial report of unruly chaos reigning out here is not quite accurate. We have now had a chance to savour other moods. Today, long slow swells, lightly cresting and spaced far apart, run smoothly ahead into the bright horizon. A gentle northerly breeze carries Porpoise along at a steady pace with the wind off the quarter and all four sails set in uniform trim to fully utilize every inch of canvas. It is exhilarating that even the slightest wind empowers our craft and sends us purposefully on our way.

Those idyllic imaginations we entertained at the thought of sailing across an ocean are realized as we settle into a more comfortable rhythm in our life offshore. We remain aware how quickly it can all change and become a nightmare, but those kinds of thoughts drift off behind us as we are lulled by a pleasant passage over this vast blue plain. Time becomes increasingly irrelevant as the days blend one into another and although our destination of the Hawaiian Islands is still anticipated, it increasingly doesn't seem to matter when we get there. Maybe we will just sail and sail and sail over these eternal swells. Maybe the journey never started and maybe it never ends.

We continued on in this way until the wind softened to a faint ripple over the rollers and then, to our wonder, died. Swaying and bobbing in the calm, we became aware of how silent, lonely, and isolated it can feel out here. Almost not wanting to break the spell, we reluctantly fired up the faithful diesel to initiate our first taste of ocean motor sailing, jostling along among the humps and hollows of the seas that seemed to have lost any purposeful direction. The sky, like the water, became bland, passing from fresh blue to dull grey which was a match for our mood as we were reduced to dogged diesel progress. It was hard to imagine what it

Cast Off Your Bow Lines

would have been like in the old days of sail to simply roll about and pray for wind. They had no option of firing up a motor—it was wait and wallow.

Two nights ago, sitting on watch alone about 3:00 a.m., I experienced a strange visitation. In the dark I could hear a chirruping sound suspended somewhere in the void behind me. It was an odd kind of sound I could not identify and seemed out of place in my lonely vigil. I thought it might be the chuckling noise of dolphins following behind in our wake. The sound ceased for a time until I forgot about it, but then came again with an unmistakable cheeping close to my head, first by one ear and then the other. Obviously this was no dolphin. The sound continued intermittently for close to half an hour until, with an abrupt slapping and flapping, the chirper landed noisily somewhere down by my feet. Intrigued and somewhat alarmed, I snapped on my headlamp and shone it down on the cockpit floor. In the ghostly white circle of light the inexplicable mystery was solved. My first impression was that a large bat had boarded us. On closer inspection it turned out to be a lovely bird, much like a swallow only bigger, with a forked tail, grey plumage, and a white triangular marking across its back. Fascinated, I watched him flutter about my feet for some time, wondering why such a wild creature would choose to come aboard this abode of humans. I thought at first it might be a lost juvenile in distress. Maybe he had wandered too far or been blown off the coast where, like Noah's dove, he found no place to land. Such an explanation was hardly plausible as we were hundreds of miles from land. Then beneath the disarray of feathers, I spied his little black webbed feet. I spoke softly so as not to startle him. "So, my little friend, you are properly equipped for a sojourn out here among the waves. You could rest on the water if you wanted, but you chose to land on our solid timbers. Perhaps you came to visit? Can we be of any assistance?"

Reaching down with both hands, I gently gathered him up and, to my surprise, he showed no fear or resistance. In fact, he nestled into the warmth of my hands and remained there, quite satisfied with himself, closing his dark little eyes as I stroked his head with my thumb. Thrilled at his complete acceptance of me, I murmured soothingly to him and made light conversation. "So…what's it like living out here on the ocean? You get cold… tired…hungry? Talk about the vagrant gypsy life. Don't you ever miss the land?"

He clearly was appreciating a break from the wind and water. I studied him closely. A small protuberance grew up from the top of his beak and curled forward giving his rounded little head the appearance of a Roman helmet with a nose guard. Later, looking through our bird books, I discovered he was a storm

110

There Be Treasure

petrel, a bird that lives at sea for long periods of time. This one seemed totally unperturbed by my human presence and made me wonder if he had ever encountered a person before.

He settled into my cupped hand, hunched himself up, and went promptly to sleep. There he stayed and slept through the night, being handed to Chris when he came on watch, and then passed over to Matt three hours later. When morning dawned and the petrel awoke, Matthew decided it was time the bird got back to the business of living whatever his life was about, so he launched him into the air with both hands. Our feathered friend immediately flew off over the swells to find his fortune. However, he seemed to think he'd found a good setup with us because the next night he was back and the next for his now accustomed routine of cuddling and comfort through the night. I wondered if all petrels are this way or if this fellow had a peculiar penchant. I hadn't heard that this was common, and it would seem to me quite a feat for him to find us again each night as we travelled up to 150 miles a day. I was disappointed when he ceased to show up and kept looking for him for a few nights afterward. Maybe petrels use boats all the time as hotels, but I would like to think he was an anomaly and picked us out special. We will keep an eye out for him.

While we are on the subject of charming ocean encounters, I have one more to relate. Yesterday Matt spied a green object afloat off the port quarter. "Hey, Pops! Look at that! Doesn't that look like green glass over there?" Straining to look in the direction he pointed, I could see something that appeared out of the ordinary, something worth investigating. "It's probably just some piece of floating garbage, but we can take a look. It's not like we have a whole lot of other things to do." We swung the helm over to check it out. As we came close, we found ourselves looking down intently at a roundish green object, its shape distorted along the water line by a strange white and grey girdle. Scooping it with a dip net, we brought it aboard and laid it on the deck. I was stupefied. It was a gigantic glass fish float as big as a basketball. The deceptive girdle was a growth of goose barnacles, and judging by the great flock of them, our ball had been adrift a long time. I related earlier that when I was a boy, we searched the Olympic Peninsula beaches for Japanese glass floats and rarely ever found them. Well, how much less likely would it be to find one floating by on the trackless ocean? So unlikely that we had not even thought to look for one. Smaller ones still occasionally wash up on the coast of North America, but one of this size would have little chance of evading reefs and rocks to be found in one piece on a west coast beach. Besides, we heard they switched to synthetic floats all the way

Cast Off Your Bow Lines

back in the sixties. Lyza was effusing over it with her characteristic enthusiasm. Matt, ever quick to seize an opportunity, swung the float to advantage. "Happy belated Mother's Day, Mom. I am giving it to you." "Aw, Matty," she said and beamed with pleasure at this perfect present. "I couldn't imagine a better gift."

What a treasure! What a place to find it! I would have given my eyeteeth for a glass float like that in the days of my youth when I searched for them on those wild Washington ocean beaches. Then years later we stumble across this one just floating nonchalantly across our track. It was like finding the proverbial needle in a haystack or perhaps more like finding an emerald gleaming in the sun on the sands of a desert. Such a find confirmed the unpredictability of locating and securing life's wondrous gifts. We cannot always find a correlation between the sweat of our efforts and the rewards or even punishments we might reap. I think maybe we just travel along day by day as we sail over this ocean of life to see what it might serve up, sometimes good or sometimes bad.

*Note added later—In our case the inscrutable work of providence, of serendipity, or whatever you might call it, was at work in multiplied measure. A few days after this amazing find, Lyza, glancing up from the dishes to gaze upon the sparkling seas, spotted a glint of green glass bobbing on a swell a hundred yards off the beam. Alerting us with a shout from the galley, we quickly turned upwind and hove to, fetching up another green glass ball identical to the first. Lyza naturally returned the favour and gifted it to Matt. Almost a week later we came upon a third exactly like the others. Finding one was amazing, two was astounding, but three seemed a miracle.

It was a reminder that the Creator more than merely sustains us; He also takes pleasure in surprising and delighting us with wonderful little coincidences to confirm with a twinkle in his eye His loving care toward us. Perhaps we should be recognizing with the same surprised amazement that the sun comes up every day. We take for granted these common occurrences, but they are a greater wonder than finding three glass fish floats on the high seas in just over a week. Even old salt Jack was impressed. In all his ocean passages he had never even seen one glass ball. He deferred a chance to own one of our finds, so this one went to Chris. How special that each family member on board will have its own memento to remember these precious days, a palantir of the Pacific placed on some special shelf at home where one can periodically gaze into its green depths and remember our common sojourn upon the sea.

Such are the wonders we are discovering every day out here where the simple offerings of nature are our daily entertainment and contemplation. Storing

There Be Treasure

up all these treasures and memories in our hearts along with the three floats we have packed away in boxes stuffed with padding to protect them, we continue on our course to the fabled islands ahead. I remember somewhere seeing copies of ancient maps where the unknown areas beyond human habitation had serpent-like creatures drawn on them with the ominous words, "There be dragons" inscribed around the elaborate illustrations. I sincerely hope we will not have to encounter any dragons out here, but with our latest experiences, we could happily adjust that saying in the margin of our Pacific chart, "There be treasure."

Yep. I think with that thought it is probably time to bring this entry to a close. 'Til next time....

Cast Off Your Bow Lines

PPSS

"Not all treasure is silver and gold, mate."

Quote by Jack Sparrow

The first of three Japanese floats

B. Smilin' Jack

May 17

Wow, time just flows like a river of consciousness out here, one day blending seamlessly into the next, the same blue view and ocean motion greeting you every morning. It's been three days since I last reported. A lot has happened out here in that amount of time. It's the tenth day on the ocean, putting us almost halfway to Hilo. Halfway to Hilo! Hmm ... could be a catchy journal title. But no, I'm calling this one "Smilin' Jack," a title you'll understand better by the time I'm finished.

The other evening the sun had set, the sky was darkening, and looking up from the helm at the lovely orange and pink brushstrokes of sunset, I was struck by the sight of a jet stream arcing across the sky. My first thought was, *What a sensible and efficient way to get to Hawaii. Let's see ... six hours watching movies while beautiful flight attendants ply you with food and drink, or twenty-four days of rocking and rolling at a snail's pace on Porpoise. Let's not even mention the cost differential of an airline ticket versus the outlay for this voyage.*

Forgive me. It was only a momentary lapse. It's all good. I repeat to myself our ocean-going mantra: "It's not the destination ... not the destination. It's the journey ... enjoying the journey." As the sun leaves us after its final exit in a blaze of glory, the heavens quickly drop a dark veil of solitude, and I know our mantra is true. This slow boat to China is our choice. I wouldn't exchange this experience for anything.

We've almost finished travelling south through the forty latitudes, and the increasing warmth is a harbinger of the blessed thirties to come, where we eagerly anticipate the legendary trade winds. Reportedly, they will whisk us to the islands of paradise on one big sleigh ride ... at least that's what Jack tells us, and already we feel some effects of the North Pacific High with clearer skies and steadier

Cast Off Your Bow Lines

winds. We're about as far south as San Francisco and 1,400 miles from the Big Island. At this rate, we may get there in better time than expected.

I don't like the way Jack keeps saying we're a lucky ship, because it sounds like he knows more about misery that we don't. To my mind, there have been miseries enough without him smirking in the background about real weather and calling waves that I think are enormous "just small stuff."

By now you'd probably like to know just who this Jack fellow is and where he came from. As I've mentioned this crew member often, I suppose it's time to give you some account of him. Simply put, Jack is our saviour. Yes, that's right—saviour. I'm not being sacrilegious or anything. Jack himself is a good Dutch Reformed salt of the sea descendant of the Reformation and all that, so he's got his perspectives clear. Nevertheless, he has saved our butts in many ways—if not eternally, then at least significantly in this world for sure.

If we hadn't listened to his sea-sage advice back on the Blaine dock, we probably would have had a mutiny on the *Porpy* by now, jumped overboard, or found ourselves floating aimlessly in the Pacific, because the crew would have been too wiped out to man the ship. What could be of such crucial importance to our survival? Well, in a few words ... lee cloths, helmsman seats, restraining belts in the galley, staysails, reefed mains, dodgers, bunk boards, noon sights, night watch schedules, shipboard routines ... I could go on, but that gives you some idea.

We've learned that every first-timer wanting to sail the open sea should have a Smilin' Jack aboard whose job it is to just smile knowingly and, with a touch of condescension in his voice, respond patiently to a multitude of ignorant questions about all aspects of this strange life. He's kind of like someone who has been to the other side and come back to tell us of things our eyes have not seen or our ears heard.

Jack has been there and back six times ... or is it seven? Even he can't remember. Anyone who would cross the same bit of ocean that many times either has knocked a few screws loose out there or has truly travelled beyond the natural to taste the sublime. Jack prefers sailing on the ocean to arriving in the Hawaiian Islands to partake of all the delights of paradise. He's going to fly right home as soon as we arrive at Hilo. Obviously the cut of Jack's jib is different from most. He really believes it's the journey, not the destination. The rest of us are all raring to hit the beaches of tropical terra firma.

Lyza and I think maybe this trip is an awesome one-timer for us. You know, it was amazing and sublime, but we're not about to trade our coastal sailing for

116

Smilin' Jack

open ocean wandering. Doing it once is enough to purchase a certain amount of notoriety for yourself and gives you enough experience to tell some good stories or, in my case, supply some sermon fodder. Doing it a second time is testing fate, as you're not likely to indefinitely avoid disastrous equipment failures, violent storms, accidents at sea, and all manner of ordeals that may be exciting to the onlooker but not for the participant. All the great sea stories you read are made interesting by the many scary and horrible things that happen to the sailors. I trust our journeying is interesting enough for you without such harrowing drama. I pray that the Lord just gets us across without too much misadventure. I figure my prayers will only hold strong for one relatively benign crossing. After that, we'd be tempting the odds.

Apparently Jack has tempted fate on a number of occasions and lived to tell the tale, and I guess there are others like him who become addicted offshore sailors, but I don't think that life is for us. So you can see why I think Jack is some kind holy ocean man or saint of the seas. Either he's a little crazy or he has tapped into something that escapes the rest of us. It's probably both, and that describes a lot of the holy men I know. Before heading out, I asked him why he would consent to come with us when he'd already done it so many times. He just said with a far-off look, "I just love it out there." Okay. Well, I've been "out there" now too, and all I can say is I'm looking at him with some amount of awe.

Jack owns the boat moored across from our slip in the Blaine marina. He agreed to come with us for the reason given above and because he kind of likes us, I guess. He especially likes "the boys," as he calls them, who are well into their thirties. He also calls them "young people." He likes the thought of "teachin' and trainin' 'em."

He didn't mention teachin' me anything, although I knew he would. I guess I'm no longer in the young people category. He refused to take any money for functioning as our mentor for the first leg of the trip, apart from the cost of his plane ticket back, and he put in a lot of work helping prepare us for the passage. I've checked to see if he has any wings tucked up under his heavy weather gear but didn't find any, and he doesn't glow at night, so I guess he's not really from the other side. Besides, I don't think an angel would drink as much rum and coke at happy hour as he does. As I said, he's going to fly away when we reach the Islands, and we'll be wondering if he really was with us or not as we head back out across the North Pacific to Alaska by ourselves.

I think God must have known he'd better send us help when He saw we were set on doing this trip. He sent us Jack, a sort of seventy-year-old Clarence

117

Cast Off Your Bow Lines

Oddbody of the seas. I'm hoping he doesn't follow Clarence's lead by jumping overboard so we'll have to save him. I hope that from here on out when you hear of Jack, you'll know I'm talking about our guardian angel out here, not Jack Sparrow or your run-of–the-mill jack tar or Jack of *Titanic* fame. Who can forget Rose crying out, "Jack! Jack!" with the freezing water gurgling about her waist as the ship was going down? No, he's just old smilin' Jack Vanderpol, the happy Dutchman from Abbotsford who jumped aboard *Porpoise* for our maiden voyage across the big waters.

Let me give you an example of Jack's worth by taking you back a few days to the twenty-four-hour stretch of our least amount of progress, and you'll also understand why I haven't reported for a while. It will also give you a taste of how variable the moods of Mother Ocean can be.

The wind had abandoned us, and we'd been motoring along at a steady pace for almost two days over calm grey seas. On the coast the wind blows one way and then another, playing off the coastal topography and temperatures so that breezes vary significantly from day to day or even hour to hour. However, on the open seas, once the sails are set, the sail trim often stays unchanged for days. But in times of transition from one weather pattern to another, there can be a few days of light or contrary winds coming and going from different directions, so firing up the engine can be helpful.

We were motoring along, getting our batteries charged up, when a hefty breeze kicked up. Finally, we had a wind, but there was one problem—the wind came from dead ahead, out of the southwest, blowing directly opposite to the direction we were trying to head. It gave every indication of setting against us for the foreseeable future. We were glad to put the sails back up after the calm, but to make any headway in our chosen direction, we'd have to go to windward and zig zag back and forth every few miles. It can be demoralizing beating along toward San Francisco when you want to be headed for Hawaii.

We always knew it was a matter of time before we encountered a low tracking northward, so we prayed for it to be a mild one and not the kind of monster Jack had met on his maiden voyage. I turned to Lyza as the wind stiffened from the south.

"It looks like we might be in for a bit of a wo pwessure awea. Uh uh uh uh huh." I was trying to lighten the mood as we faced our first real spate of opposition.

The joke referred to an announcer on our VHF radio who had a total Elmer Fudd voice. He regularly came on air one summer over the coastal weather station: "We will be having a vewy wainy weekend in your awea because of a wo

118

Smilin' Jack

pwessure depwession in your wegion. The winds will be vewy stwong in Georgia Stwait." Lyza's brother, Chris, who was with us on *Porpoise* at that time, thought it hilarious, setting us to making our own imitation Fudd broadcasts at the least pwovocation. Lyza, however, was clearly not in the mood for any of my "wo pwessure" jokes, as she assessed that the wind was steadily increasing.

Jack turned his gaze to the horizon and then up to the mainsail. He gave us his usual grimace and broke into a broad grin. "I guess you'll find out why I hate going to windward," he chuckled.

We discovered that going hard on the wind out on the ocean means that you pull in all your sails as tight to the ship as you can by hardening the sheets and point your nose as close to thirty degrees off the wind as possible. The ship heels sharply over so that everyone has to brace against the steep angle while your boat bashes against the big seas that splash up over the bow and swash over the deck. The wind bears down sharply as you proceed in this direction in which you don't really want to go. As you persevere, there is nothing to do but hang on grimly and hope to God the wind veers, fades, or dies before you do.

The only alternative to this grind is to "heave to," which doesn't mean you finally throw up. To heave to, or to be "hove to," which you've probably read about in any good seafaring novel, but didn't understand, means you backwind the foresail, throw your rudder up into the wind, and ease the main sheet. This manoeuvre puts you into neutral and essentially parks your boat at about a forty-five-degree angle to the waves, greatly easing your passage. Cool, huh? The downside is that although you're vastly more comfortable, you make no progress toward your destination.

In this instance, we concluded that the conditions weren't bad enough to heave to, so we bashed along in this hard-to-windward manner for three or four hours. At that point, I suggested to Jack that maybe we ought to heave to.

"Weeelll ... yeah ... we could ... but it would be a shame to lose the two or three knots we're getting."

I decided not to point out to him that those knots were in the wrong direction. Jack's an old salt, and I didn't want to be a wuss (you know— a wussie—one who can't take it), so I just murmured, " Yeah, right, we wouldn't want to lose those knots." I actually thought maybe losing the knots in our stomachs was enough compensation for losing the nautical mileage per hour, but I kept silent. A nautical mile, by the way, is equal to a minute of latitude, so it's longer than a land mile, calculated as it is on the size of the globe. A nautical mile is around 6,076 feet, while a land mile is 5,280 feet. Just so you know.

Cast Off Your Bow Lines

We carried on stoically into the wet, cold night. I was on the first watch and perched myself on the helmsman seat, bracing myself against the thirty-degree heel of the ship. A dark tumult of threatening clouds moved swiftly across my field of vision in the dim light of the moon. Soon a pelting downpour announced itself with a rattling attack on the plastic dodger window. It battered us for almost an hour, and then suddenly the wind veered, lessened a bit, and seemed confused about what to do next. I came off watch while it was still making up its mind, turned the helm over to Matt, and thankfully went below to peel off my wet clothes, crawl over Lyza's sleeping form, and collapse in my bunk. Sleep instantly took me, and I knew nothing more until I awoke to a great clatter and the sound of loud voices above me.

"Tie it off! No! Not there over here!"

Whack! Crash! Flap! Flap! Flap!

"Oh (expletive deleted)!"

"Look out!"

Crash! (The boat heels way over to port.)

"We gybed!! Matt, you've gotta take the mizzen down."

"I can't! It's pinned. It won't come down!"

"It'll come—just yard on it. Yank it down!"

Being woken from a dead sleep, I was in a fog of incomprehension as I tried to grasp where I was and what was going on. I dimly recognized Matt's voice and then Jack's above the noise of wind, waves, and straining gear. As the boat rolled sharply, I felt my body pressed against the hull as I tried to take in the situation. The weird thought that came to me was, *Well, we're sure as heck not in Kansas anymore, Dorothy.* Rather than safe at home in my bed, we were in the middle of a gosh darn huge Pacific Ocean in crazy, chaotic seas.

I considered getting up to help out, but the noble resolve quickly receded. It was so much easier to pretend it wasn't happening. I feigned sleep until a familiar voice spoke in the dark.

"You're not really sleeping through this, are you?" I remained silent, but it didn't fool her. "John, do you hear what's going on up there?"

"Uh ... kinda."

"Sounds like they need help."

I sheepishly remained immobile, as though I was just now noticing the mayhem on deck. The shouts above grew more frantic. I began to contemplate what it would take to get past the dividing board between us and crawl over Lyza and the lee cloth in order to stand on the extremely tilted floor, shivering in my

Smilin' Jack

boxers while donning my clothes, my heavy weather gear, harness, tether, and life vest. Easy for my wife to volunteer me. I didn't see her jumping out of bed to help. I pulled the covers over my head, but immediately my conscience began to smite me, and I finally made as if to rise, only to fall helplessly back again.

"John! Aren't you going to go help them?"

"I think they've got it under control," I said, but what I heard was …

"Jack! Jack! The sail is stuck halfway!"

Flap! Slap! Whack!

"Where is that stupid wind coming from now? I can't see a thing. Is it veering north? I think it's veering north. We've got to get the staysail down first."

The minutes passed with fewer frantic-sounding shouts back and forth. My reluctance to go above was overpowering, and in the end, my procrastination paid off. I was relieved when I heard, "Okay, Matt, I think we got her. That's good. Keep her steady on that heading."

The boat seemed to be riding more smoothly, and the wind's sound and fury was subsiding. The bedlam on deck had sorted itself out too once the direction of the wind was determined and appropriate adjustments were made—all without my help. I don't think I was needed after all … at least that's my story, and I'm sticking to it. I could still feel some reproachful vibes from the other side of the partition, but I'm sure that would sort itself out in the light of day … or maybe some other day … as long as it wasn't now. I slunk back under my covers and returned to fitful dozing. My turn to handle a crisis up there would come soon enough. In the meantime, no worries. Smilin' Jack was on the job.

Cast Off Your Bow Lines

PPSS

"There must be some easier way to get my wings."
    Quote from Clarence Oddbody in *It's a Wonderful Life*

Smilin' Jack and Lyza

14. Catch the Trade Winds

May 20

Angel Oddbody on the job or not, we've made it several more days, and in that time we've noticed a big change. We have entered into the thirty latitudes where both air and sea are warm enough that we shed our northerly heavy weather gear and go about in shorts and tee-shirts, even in the night watches. The vicissitudes of the earlier part of our voyage appear to be over. The northeast trades strengthen each day off our starboard quarter, and our sails are set to take advantage of this broad reach point of sail, needing very little attention to the trim day after day in the consistent breeze. Jack says it will probably be this way all the way to the Islands.

Now you can be legitimately jealous. Picture in your minds the most idyllic scene you can imagine. By this time we're used to living with the ocean motion, and with seasickness behind us, we're literally catching the trade winds in our sails. It's truly the sleigh ride people say it is. *Porpoise* positively exults in the wind and waves, galloping along like a thoroughbred let out of its stall on a fine day. With fifteen to twenty knots filling all sails, I swear I could actually feel the joy radiating through her trembling timbers. I couldn't help but think of Kevin, her mechanic, during the journey preparation stages, cautioning us about taking her offshore.

"Well, you know these Hudson Force Fifties were never really made to be offshore boats. Besides, she's getting old. There could be many problems."

If only Kevin could see her now! He was a meticulous mechanic and more of a powerboat kind of guy, so maybe he just wouldn't get this, but she was made for sailing offshore. The way of a ship on the sea is, after all, one of the four great mysteries, according to the Bible. I settled in the helm seat and patted the old girl affectionately on her binnacle.

Cast Off Your Bow Lines

"Don't you listen to him, *Porpy*. You've got what it takes. Not an offshore boat indeed!" I thought I heard a snort as she quickened her pace under a fresh gust and surged forward.

Even Jack has only compliments for the way she handles as a bluewater boat.

"She's balanced and steady and moves right along well in the seas. She doesn't wallow and keeps a straight course. But then … she's a lucky ship."

Yesterday we figured we reached the halfway point. Gathering the crew, I popped a champagne cork far out over the water. Pouring drinks all around, we toasted our passage, shouting, "Hawaii or bust!" with a brut called Scharffenberger that soon became Scharffenbarfer in defiance of the seasickness we left behind.

The "boys," as Jack calls them, got us swimming when they begged for a dive into the mild turquoise water yesterday. It's intimidating to actually consider leaving one's only safety and jump into the ocean when you're a thousand miles from any land and the bottom is five miles down, so we had to think it out carefully. We waited for a relatively calm moment and then brought *Porpoise* up into the wind and hove to, setting down the boarding ladder. The plan was for the boys to perch on the bowsprit, dive one by one into the sea, pop up quickly, and grab the ladder as the boat drifted by. I stood by with life ring and heaving line in case anyone missed his grab. It gave me the willies to imagine one of the boys drifting off astern and being swallowed up in the blue expanse. They are young and agile, however, and made the operation look easy.

After they set the pace, we all eventually overcame our trepidation. One by one, the whole crew, even seventy-two-year-old Jack, made the mad leap into the exhilarating refreshment of a saltwater bath. To our surprise when we opened our eyes underwater, instead of a fearful dark abyss we experienced the warmth and aqua blue of a giant swimming pool. Our customary northern seas are greenish in colour and emanate a mysterious feeling of dark and deep when you peer into them. This tropical ocean, although far darker and deeper, felt almost friendly because of its warmth and colour.

Another fascinating development in our warming world is the increasing appearance of flying fish. We had often read about them, but it's altogether different to actually see them for the first time popping out of the waves and skittering off ahead of the bow. Even as I write this journal entry, I glance at the windward side and see a launch of what I've come to call a "flock" of these creatures, as they always seem to spook in groups. These little fish are amazing but I think somewhat defective in one crucial aspect. They spring up suddenly, wings fluttering furiously like dragonflies to skim over the waves for thirty, forty, and occasionally

Catch the Trade Winds

hundreds of feet, sparkling and shimmering like so many Tinker Bells. Then just when you've come to admire the wonder of their ability to fly (They're fish, for heaven's sake!), they manifest their one defect by dropping abruptly back into the sea in a most clumsy and undignified fashion. It's as if suddenly they realize that fish aren't supposed to fly and, like Peter walking on the water, lose faith.

"Aaaahhhh … look at this! Look at me! I'm flying! I'm a fish and I'm flying! Yikes! Will ya look at those waves! Abort mission! Abort!"

Kersploosh!

It appears God—or evolution, if you're so inclined, which I am not—forgot to equip them with skidders to ease the re-entry procedure. You'd think they'd have pontoons like a float plane, or at least webbed feet to slide in on the surface like ducks on a lake. Perhaps they would think themselves just too cool if this deficiency were corrected, so they were left without to keep them humble. On the positive side, this stop/flop manoeuvre and the resulting sudden crash of their descent may be designed to make hungry dorados waiting underneath work harder to catch them. It would be easier to time their chomp at the end of a slide trail. Humility has its benefits.

You may be wondering whether I have too much time on my hands after reading my flying fish account or conclude that boredom must be a problem out here with nothing to do on our small craft while we sail and sail and sail upon leagues of unchanging scenery. I trust that some of the entertainment I have described, as well as my descriptions of how much effort it takes to do the simplest chores of living, partly answers that question. It's true that there are many hours to fill, so I'll let you in on some of our other diversions.

One occupation is to watch for floating debris. There's a lot of flotsam out here. Sadly, most of it is plastic. This crap lives here forever. It's not wall-to-wall, but every once in a while you see something that's not organic, such as plastic fish floats, synthetic fibre nets, ropes, containers, and plastic bottles of every description that have been bobbing around out here for decades and will till Jesus comes. Plastic lives much longer than we will. Let this serve as our environmental awareness call-to-arms.

Jack thinks it's okay to throw paper products overboard because it all decomposes. "Concentrated fibre," he calls it. Anything organic from the table is fine to toss, and flushing the heads is no problem out here. Jack has some fairly crusty politically incorrect views about that business.

125

Cast Off Your Bow Lines

"I'll start worrying about flushing the toilets when whales start wearing diapers," he says with disdain. Something to think about. I imagine one whale poo would more than quintuple our output for the whole voyage.

We are being as environmentally sensitive as possible out here. In the bad old days of my childhood, my mother used to cast old items over the side of the boat on our family vacations while singing "Nearer My God to Thee," striking a sombre note of finality for the departure of empty cans and bottles. You knew you were never going to see that beat up old hibachi again once it was dropped over the side. On our voyage now, cans and glass bottles get the deep six this way (we dispense with the singing), as long as they sink. They descend the five miles or so to become homes for destitute crabs or other sea critters. No one will ever get down there to see them, and it's pitch black anyway. God can see, but I don't think He cares much, as His final incineration plan will dispose of the whole lot of everything.

Such metaphysical contemplations are common out here with all the time we have to meditate. We contemplate things like how long it would take a bottle cap to hit bottom. We figure five miles would take a couple of weeks at least, but no one has done a scientific calculation. We're too lazy for that, just like we're a too lazy to get on top of sun sight calculations. Lyza went to the trouble of taking complicated classes on using a sextant, but we always intended to have that ancient instrument as a backup if our electronics got zapped by a lightning strike or something. It would take that or something equally serious to stop relying on our chart plotter. Who wants to do all that math when you can just follow the little pink road the satellites give you? But we do make daily bets on how many miles we have travelled each day (which the chart plotter tells us) or how many sail changes Jack will want to initiate in one afternoon. Am I answering your questions about how we spend the time out here?

Another oddity that happens is that we start to anthropomorphize some of our inanimate sailing gear. This could be due to a desire to create new crew members or just because we're getting a little silly. The autopilot has naturally become Otto, and naming him is probably most appropriate, since he takes regular crew shifts on the helm. He's not a bad seaman either, guiding us through many a dark night on a straight course.

Otto is very reasonable in calm coastal sailing waters, but out here he's become a little neurotic. He's taken to stuttering and making wild, fearful course corrections under the slightest provocations. His gear consists of an electronic compass, a reversible pump for the hydraulic steering, and an obsolete Wagner

Catch the Trade Winds

electronic brain. We can all hear this gear below decks when we turn him on, sounding a little like an indecisive bumble bee in a flower patch.

Mmm … mmm … uhn, uhn, uhn … mmmmmmm … aheh aheh aheh Aaaarrrrrrrrrrrrrhhhhhg … mmm mmm mmm (a momentary silence) *oorrrrrr-rrrrrrraaaaaach.* The big seas naturally push and pull the bow of the ship quite a bit, which Otto always takes personally as an affront to his capabilities, hence his wild and frantic adjustments. Sometimes Otto starts screaming at the top of his mechanized lungs, and we try to reason with him: "Hey, calm down, buddy, it's not that bad. Just relax a bit and let the bow come back to your heading on its own."

If he's too upset and we can't persuade him to listen, we give him a time out and hand steer in order to save our nerves and his.

We also have George, our man overboard marker. We've thrown him over a couple of times and then rescued him. Better George than Matt.

Then there's Genny, the Genoa sail, a good ol' girl … you get the idea. Someone thought they caught sight of Wilson the other day. Yes, that's right, Wilson from *Castaway* fame, floating right by us, all bedraggled and faded with bamboo shoots stuck through his head. By the time we thought to rescue him, it was too late, so we noted our position of lat and long and thought we'd mail it to Tom. "Sorry, Wilson. Wiillsooon! We're sorry, Wilson."

Too much time on our hands perhaps? Yes, it can get a bit weird day after day with nothing but water, water everywhere. If we all arrive in Hilo talking to the halyards, someone call the authorities and get us some psychiatric help, okay? I notice lately that Jack's been cozying up more to his rum bottle. If he starts calling it by a name, we may have to do some sort of intervention, or at least suspend him from night watches. I'm keeping an eye on Chris as well after I saw him wrapping up his coveted glass ball and hiding it under the salon bed. He said it was just for protection, but if he starts calling it Julie, we'll know it's time to put a stop to this anthropomorphizing thing once and for all. Actually, our lives are much more disciplined than I'm making it sound here. Jack has taught us some seaman's routines to keep us and the boat shipshape and in Bristol fashion. More about that next time.

127

Cast Off Your Bow Lines

PPSS

"The fair breeze blew, the white foam flew,
The furrow followed free;
We were the first that ever burst
Into that silent sea."

Excerpt from "The Rime of the Ancient Mariner"
By Samuel Taylor Coleridge

Matt catching the trades

15. Fiddlin'

May 24

Our existence may seem aimless or random to those of you following us from an armchair on terra firma, but life on an ocean-going vessel does conform to certain disciplined routines and observances. We wouldn't have been aware of such an ordered existence at sea if it weren't for Smilin' Jack. He quietly did his thing until a discernable pattern emerged that, through his fatherly influence, we adhered to. It goes something like this.

First you get up. Well, that might sound obvious, but there are definitely some mornings when I don't feel like confronting those chaotic seas. If not called to a watch, I linger as long as possible until, forced by the urgencies of nature, I crawl out of my tiny cubicle. More on that later, but suffice it to say it requires a lot of effort to extricate myself from the covers and leverage myself over Lyza to gain a standing posture on the heaving floor.

Bracing myself with one hand on each wall, clothed only in my boxer shorts, I peer out through the small rectangle porthole of our stateroom at the insistent sea slithering and foaming toward me and mutter something like, "Will these wretched waves never stop?"

My attitude at times settles into a grim malaise after bad weather or late sleepy watches because of a form of seasickness that has grown upon me as the days at sea rack up. I haven't experienced the Juan de Puca kind that Chris was afflicted with early in the trip, but I have developed a mal de mer of a different sort, characterized by a light-headed irritability, a slight aversion to food or alcohol, and an overall sensation that my brains have been scrambled. This settled in after about a week and half and has stubbornly remained. Everyone else, including Chris, are quite chipper in the mornings, so I feel it my duty to mask my symptoms and get with the program.

Cast Off Your Bow Lines

Once up, the most important next step is coffee time. Sucking back the java each morning is every bit as enjoyable a day starter for me as it is for you drinking yours as you read this entry, making a nice bond between us all, even across the miles of ocean. There's nothing spectacular about the coffee time process other than the usual gymnastics required to accomplish anything on this rocking boat, which is why we obtained a stainless-steel French press to prevent the calamity of losing our brew. The other day, Lyza, now a sea veteran of a couple weeks, was a little smarty pants and made for the table with sugar in one hand and cream in the other. A wave hit, and she flew across the room, striking her upper arm hard against the pots and pans cupboard. Thankfully, the arm wasn't broken, although the bruise is nasty to look at.

After coffee, we put together some breakfast, a business lasting until at least ten-thirty. Then each crew member greets the new day in his or her own way, meditating on the sea and the endless horizon, inhaling deep drafts of fresh ocean air, and scanning the waves for any interesting activity. Watches are more casual in the day, with most of us cheerfully ready to take the wheel, especially because the helmsman's chair is the most comfortable seat in the house. Often when skippering, the boys strum their guitars, steering with their feet, and hum along with the rhythm of the wind in the rigging.

It's not long before the most important hour of the day arrives at the stroke of noon, the beginning of each twenty-four-hour seaman's day. That's when the sun has reached her zenith and we can (but don't too often) take our sun sight and confirm our position with the GPS. After we pinpoint our latitude and longitude on our big ocean chart, we draw a line from the previous day's position and calculate the number of nautical miles travelled. The GPS tells us all this information at a glance at any time, but we like following this noon ritual as mariners have done for ages.

At the noon mark, before anyone looks at the instruments, each of us guesses how far we travelled in the previous twenty-four hours. This requires thinking through how fast we were going, the wind speed, how many times we had to alter our course from the straight line, and whether somebody fell asleep on watch the previous night and awoke after who knows how long to find he was heading the boat back to Canada. Such inside information can give a real advantage, except you don't know if the other crew members hold similar secrets. After each one enters his or her guess, we check the chart plotter, and the winner earns a round of applause and gets out of that day's lunch dish duty. Our worst day of progress was fifty-four miles when we found ourselves going to windward, bashing about

130

Fiddlin'

in the wrong direction. Our best day so far is a sweet 162 miles in these cooper-
ative trade winds.

Once we're done with that ritual and have duly recorded it in our log book,
including weather comments and daily anecdotes, Jack usually proclaims it ought
to be Miller time or Bud time or whatever-beer-we-got time as preparations for
lunch begin. Episodes of bizarre humour can arise at any time out of a giddiness
that I try to blame on a certain form of seasickness. Hilarious laughter erupts
over stuff you probably wouldn't think even merited a chuckle. Out of the blue,
someone might start singing a stupid little ditty, behaving in some goofy manner
while doing calisthenics, and breaking out innovative dance moves along with
the exercises.

One day I got a quirky inspiration for an offshore Bud commercial. It involved a crew member poking his head up from the companionway to exclaim,
"We're out of Bud!" Matt and Chris soon got into my kooky idea with their own
contributions, and before long the Air Force and Navy were flying and steaming
to our rescue in a full-scale intervention, complete with Navy seals boarding
Porpoise, and helicopters flying off aircraft carriers to lower cases of Bud onto the
deck. I'm pretty sure it could have been Super Bowl material. Moving on.

After lunch, Jack leads us on through the seaman's day to the afternoon nap.
The regularity with which Jack attends to this duty has inspired us all to follow
suit, except for the people on watch, although they probably nod off as well.

"You gotta look after yourself out here," Jack says.

Amen! Since we all lose some three hours sleep each night during our watch,
we have no problem becoming disciples of the seamen's saint in these matters.

"We gotta look after ourselves," we say. It's a good mantra when you want
to get out of something, like doing the dishes or straightening up your sleeping
quarters. "Uh, I don't think I'm feeling up to it right now. I gotta look after myself, you know." So the nap is an accepted part of the day for everyone. When
you see a person's eyes start to droop, you know that's where it's headed, or when
someone starts to lose it and starts screaming at the rollers. We gently shepherd
that person off to his or her bunk and tuck them in, hoping a little sleep will put
a little more jack tar in their vitals.

After Jack's nap time, which usually lasts about an hour, he awakes all fresh
and perky in a way that's surprising for his age. In short order, he initiates the
next phase of our day—I've started calling it "fiddlin' time." No, Jack doesn't
have a violin. That would be right up there with bringing bagpipes aboard. If he
started with that kind of practice every day, I'm sure it would quickly be vetoed

131

Cast Off Your Bow Lines

by the crew. But every afternoon about two o'clock, we observe Jack get what we call "the fiddlies." After setting a spell on the back cushions after his nap, a certain look comes over Jack's face—a Gyro Gearloose light bulb appears above his head, and I know the next step is hauling out the toolbox. Jack is about to embark on his next *Porpoise* improvement project.

He can be pretty cagey about getting the go-ahead on any substantial changes. He'll appear beside me at the helm and begin a conversation.

"Aaahhh, you know that staysail? Well, it really shouldn't be sheeted outside the shrouds."

"Really? Where should it be?"

Jack will stroke his chin and crinkle his eyes up. "Weelll, you know those rings on the deck by the main mast? Those are for the staysails."

"Oh really? I wondered why those rings were there. I thought they were for the whisker pole or something."

Jack will give me a pitying look over the rims of his sunglasses and say, "Nope those are for the staysail. Now here's how we could rig it up ..."

Pretty soon, as long as he gets my okay, off he'll go fiddlin'. Believe me, just about everything has been fiddled with and brought up to snuff for Jack's standard of going-to-sea readiness.

This particular fix involved re-rigging the staysail sheets to run through a block attached to a brass ring already attached to the deck so that we could draw the sail closer to the centre of the ship. That sounded like a good idea to me, so I signed off on it. The breeze was quite brisk just after we did it, and I could see a lot of tension on that block and ring, which worried me a little. Sure enough, about an hour later, the sail started suddenly flogging, and I ran up to investigate. To my dismay, I discovered that the ring had pulled right out of the deck, three-inch screws and all.

"Jack!" I yelled. "The ring just yanked right out of the deck. What are we gonna do now?"

He came hustling up to check for himself.

"Waaaalll ... you see ... there really should be a proper track for the block laid in up there so you can adjust it forward and back."

"Well, that's not going to happen out here. We'd need a track; besides, that sounds like a pretty major job. Maybe we can look into it when we get to Honolulu, but this is a problem right now."

"Naw, naw, we'll just take the sail down for today and bolt the ring in tomorrow. The track can be done later."

Fiddlin'

Jack's quota of fiddlin' energy for that day had obviously been spent, and I could see his gaze beginning to wander toward the galley.

"I've got some ideas we can work on tomorrow," he continued. "I think it must be about time for happy hour, eh?" And off he ambled to find his Navy rum.

The next day, Jack appeared on deck right on time for fiddlin' hour with tools in hand. To accomplish his revision programs and get me outnumbered right off the bat, he organized an alterations crew consisting of the boys and himself. He's made it his personal assignment to teach 'em and train 'em to be good sailors and fiddlers. I think he figures they are young, mouldable, and enthusiastic about his enterprises, unlike their father, Cap'n Whitebeard, who sometimes feels these projects are a bit ambitious to undertake while rolling on the ocean. Don't get me wrong. Jack is always respectful and defers to the captain of the ship, honouring the chain of command and all. But it can't hurt when you've got a strong agenda in your mind to bolster your case by enlisting the support of a crew. Off he headed toward the foredeck, followed by the boys with sheepish grins, carrying drills, hammers, wrenches, and a tube of 5200 black Sikaflex. I made faint protestations and averted my eyes as they settled down on the foredeck and proceeded to make drilling and hammering noises. I didn't want to see or know. I was dealing with irresistible forces.

When Lyza saw the boys head off with the fiddler, she decided a bit of feminine intervention was in order.

"Jack, you're not about to drill holes in the deck, are you? That's right over the forward bunk. We can't have any leaks on the beds!" Clearly this was an alarming prospect to my co-captain.

Jack didn't want any trouble from this quarter. He and Lyza had a harmonious and affectionate camaraderie. He spoke soothingly. "Oh, you don't need to worry about that. It won't be a problem."

Lyza crossed her arms and didn't look convinced. Jack knew where Lyza's vulnerabilities lay, however, and took a different tack. "I think this is a very important thing for the safety of ship and crew. The staysail has to be secure." And he added, "It won't leak. I promise."

She gave in. He had struck one of Lyza's most important sensibilities, and they both knew it. In short order, the alterations crew put two holes right into the deck through the support joist: one into the shower stall of the forward head, and the other one over the forward stateroom bunk. Jack just happened to have on board a threaded rod of metal that they custom cut, inserted through the holes, and secured below with washers and nuts. (Who carries threaded rods in

133

Cast Off Your Bow Lines

their personal luggage? I ask you) He was the old salt. We just had to put our trust in him as he sat on the deck amidst his tools, wood shavings, and blobs of black Sikaflex on the teak deck. His fix ended up working well enough and did improve the function of the staysail.

The jury would remain out on the possible leakage issue. I've gotten pretty handy myself around boats by now, but I have a real squeamishness about cutting or drilling holes through decks and timbers. It feels like a violation and a risky weakening of our defences against water penetration, always a major battle on older boats. Not Jack. You just grab the tool and go to it, drilling or cutting away.

Even when there doesn't appear to be much in need of fixing on any given day, our salty old sailor finds something to tinker with—the set of the sails, the lines, the blocks, anything, just to get the itch out of his system. The other day Jack came off his nap and said, "Aaaaahhhh … maybe we should take the mizzen down and see how she sails with just the Genny."

"Jack," I said patiently, "she's sailing just fine. My motto is if it ain't broke, don't fix it. Matt and I tried that sail change a few days ago at your suggestion, and the boat wallowed too much. The mizzen steadied her out. What's different now?"

Jack fixed that imperturbable gaze on me, saying, "It's good to try things. It's always good to try things."

I could tell he was pretty set on the idea and was already making a move to head forward to find something else to adjust. Irritated, I was about to intervene and tell him to leave the darn sails alone when a glance at the clock brought me up short. Wait a minute— two o'clock. Of course! It's fiddlin' time. Might as well be the mizzen sail as re-tooling the hydraulic system or some other drastic alteration. I let him have his head and we went through an hour of fooling with lines and shackles and sheets.

"There, I think she's running smoother now," Jack said smugly.

"Sure, Jack, sure … that's great … lots better. Hey, isn't it just about happy hour?"

I first met our Dutch sailor the previous spring back at the Blaine marina and had no idea then how large he would loom in our lives on the ocean crossing. Jack owned the sloop named *EcstaSea* in the slip across from us. How could I have known then that ecstasy for Jack would mean not only sailing boats but messing with boats? As soon as we got to know him, and after he agreed to do the first leg of our trip with us, he started turning a critical eye on *Porpoise*, freely dispensing advice. That's when we first encountered his cowboy-like drawl. It usually comes after you ask some question.

Fiddlin'

"So, Jack, what do you think about our sleeping arrangements with these bunks? Are they gonna be okay for going offshore?"

"Waaaallll … aahhhhh." A pregnant pause follows while he eyes up the situation, stroking his chin and cocking his head while you are suspended with him in grave contemplation. You can be pretty sure that whatever he's about to say, you probably haven't thought of it and undoubtedly aren't going to like it.

"You're gonna need to have lee cloths. You can't go to sea without lee cloths."

"Right, I wouldn't think of going to sea without … lee cloths." I'm thinking, *What the heck is a lee cloth?*

He carries on like I should know. Jack rummages under the mattress to check the viability of attaching them to the plywood while I'm already thinking that this idea feels like a nuisance.

Jack catches my expression and shoots me a rim shot above his glasses. "You're gonna need 'em … I'm telling ya. If you want to stay in your bunk."

"C'mon," I say, "as if you're gonna fall out of your bunk." Does he think I need bunk diapers? Anybody can manage to stay put in their bed, can't they? Later I look up the definition of lee cloths and find out they are pieces of canvas about a foot or more high, strung up along the side of your bunk, universally considered absolutely essential pieces of offshore equipment. I sheepishly set about hiring Dan, the canvas man, to make us some lee cloths. I can't help but think of loin cloths in conjunction with this new phrase but don't see any connection.

Later on, Jack went beyond preachin' to meddlin' when he fixed his attention on the double berth where the captain and his lady sleep in the stern quarters.

"Aaahh … you're gonna need a board secured in there."

"Where? What do you mean a board?"

Jack squared up to me, "You're gonna need it. Believe me. It's no fun getting tossed around into the person next to you."

"No, Jack, really, it's okay. Lyza and I sleep very close … like two spoons. I don't think we'll need a board."

He just grinned and chuckled under his breath, good humour all the way, and then slowly shook his head with an all-knowing twinkle in his eye.

"I'll just make up a little something at home. We can put a bolt right through here and maybe a lag bolt there."

I left him to his devices and later reported the exchange to my bedmate. "I don't think he really heard me when I said we didn't need it," I said.

135

Cast Off Your Bow Lines

Lyza looked a little concerned. "I hope we're not going to have trouble with this guy. Are you sure we're going to get along?" Lyza is always about the relational ramifications. I'm mostly thinking about logistics.

"Well, he's really affable, but I'm not going to take everything he says as gospel."

In this case, gospel with Jack was indeed a straight and narrow way—about twenty-two inches for Lyza and thirty-two for me. I kid you not. His plans for us became clear when he came walking up to our slip with plywood under his arm and some angle iron in the other hand.

"I think this ought to do the trick."

He clambered aboard with all his paraphernalia and went right to work, drilling holes through the finished cedar panelling above our bed. Now it was me making the grimace and wincing as the drill broke through the wall to the cockpit. I looked on helplessly as he fitted the angle iron to the plywood and mounted it with lag bolts to the wall, splitting our bed space in two. Noting with dismay the cramped narrow spaces, I started protesting.

"How the heck am I supposed to get in there, Jack?"

The seventy-two-year-old scrutinized me with a more serious appraisal. "Do we have some kind of health issues here?"

When I assured him I didn't, he said, "Then you just have to do this."

He crawled up on the bunk, thrust his head and shoulders over the plywood, and with a deft roll of his hips, flipped in the air over the board and landed on his back in my designated spot.

"You'll be just fine in here. You'll get used to it in no time, and you'll be thankful for it. Believe me."

When Jack left the boat that afternoon, I called Lyza at home and told her she better get down to the boat and take a look. After assessing the bunk alterations, she looked at me dubiously and said, "I guess we better try it out."

We crawled into our places to try them on for size. I was kind of peeved that I had a much more difficult time getting into the space than Jack did. Peeping over the top of the board from her side, Lyza smirked a little as I lay glowering in my restricted space, shoulders squeezed between hull and board. Its deficiencies were obvious to me.

"This isn't exactly an intimacy encourager, is it, darlin'?" I said.

She smiled sweetly at me. "Well, at least our heads are clear of the board, so we can still talk."

"That's not exactly the kind of intimacy I meant."

Fiddlin'

She fixed those blue eyes on me with a look of incredulity that made her position crystal and left me with no illusions about that matter. A sudden surge of assertiveness came over me.

"Okay, well listen, I respect Jack. He probably knows a lot more about all this than we do, but he doesn't know us and how well we sleep together, and I'm not putting up with this gosh darn board. I'm calling him tonight and letting him know. You have to set boundaries sometimes, and I'm putting my foot down."

Jack was nice about it on the phone and justly mildly said, "Well, it's all made up now, so how 'bout you just take it along and tuck it away somewhere in case you change your mind."

That sounded like a good compromise to me, and I figured it could slide in along the wall in the head without it being in the way.

The first night in the open sea, Lyza and I cuddled up together and drifted off to sleep. However, we had vastly underestimated the degree to which unconsciousness removes the body's natural tendency to brace itself against motion. Peaceful sleep requires total relaxation of the body's muscles. After we went to bed, the seas really kicked up. I have to admit, I have occasionally entertained fantasies of my wife throwing herself passionately at me in the middle of the night, but my come-to-me-my-darling imaginings were rudely shattered by the body slam I received just after midnight.

As I tried to shove her back toward her side of the bed, the boat lurched the other way, and I rolled heavily in her direction. Suffice it to say my normally sweet wife has never had sanctified responses to jarring events at night. As time wore on, let's just say the atmosphere between us was not what anyone would call amorous. Murderous maybe. By 3:00 a.m., Lyza was demanding I get the cursed "chastity board," as I called it, and screw it in place. Sleep had completely fled by then, and all I could muster as I resolved to get out of bed was, "I'll just find another place to crash until my watch and figure out the board tomorrow."

The next day, I dug the contraption out of the head and installed it down the centre of our bunk, murmuring, "Bless Jack, bless Jack, bless Jack, the seaman's saint" as I screwed in the lag bolts. A famous proverb says, "It is better to dwell in the corner of an attic than in a spacious house with a brawling woman." My sea adaptation is, "It's better to sleep alone in a coffin-like space in the corner of a ship than to roll about on a spacious bed with a vexed (and vexing) woman."

Sorry, I've really gotten sidetracked with my Jack stories, but as you can see, he provides endless fodder for my rambling scribblings. I started out in this journal entry to describe our daily schedule. So back to the seaman's day. After

Cast Off Your Bow Lines

waking, coffee, breakfast, the noon sight rituals, lunch, nap, and fiddlin' hour have all come and gone, the light mellows and the day passes into that time indelibly imprinted into a salty Dutchman's soul called "happy hour." This hallowed hour happens every day about four-thirty as regularly as the sun rises and sets. The sight of Jack descending the stairway toward his stash in the locker above his berth, plastic glass in hand, is the sign that the transition has arrived.

Every member of the crew was instructed to provide for their own special treats before the trip. Each person knows what candy, nuts, chips, or particular beverage he doesn't want to run out of before we reach the Islands. Jack's favourite taste treat is of the liquid variety, a ration of rum every day in the finest British marine tradition, even though he's Dutch. In his case, it's usually a little more than one tot. As soon as he emerges back up on deck with rum and coke in hand, it's a signal to the rest of us that the day is done and evening activities have begun, with anticipation of our own sundowners and the delicious dinner Lyza has concocted for us.

Jack is always circumspect with his happy hour, but there's no question he becomes more talkative as the hour progresses. Usually by the time the aromas are wafting up from the galley, Jack can be found holding court with the boys on the back seat, expounding on many subjects as the they squeeze everything they can from this mentor of the sea they've come to like and respect.

"Jack, what's the worst conditions at sea you've ever been up against?"

"Weealll ... aaahhhh ... let me see ..."

"What do you think about global warming?"

The sea sage will respond, giving room for other opinions, sometimes salting in his colourful and often politically incorrect perspectives, and sometimes spouting surprisingly progressive views on a variety of current topics. He's pretty well read and has lived enough life to temper his conservative background. The boys love getting him going.

By dinnertime, we're ready to consume generous helpings of whatever is served up. I'm talking real meals here—beef stroganoff, teriyaki chicken and rice, spaghetti and meatballs, often topped off with a homemade dessert. In spite of the rolling seas, we eat well and are burning off the calories with all the activities required to keep our boat afloat.

Jack gets quieter and quieter as dinner proceeds, and by the end of the meal, he's nodding and practically comatose, staring at us under heavy lids, his trademark grin plastered fixedly on his face until he announces that it's time to retire to his stateroom.

Fiddlin'

This is a signal to the rest of us that the first watch needs to ready himself for the climb to the helmsman's chair while the rest of us eye the pile of dishes and hope that someone else is feeling especially sacrificial so we can edge off toward our own bunks. Doing the evening dishes ranks right up there among the most heroic acts that a crew member can undertake. The chore is made more tolerable by playing music CDs, and again Jack has come through for us by bringing a great mix from the fifties and sixties. Sometimes one of the boys serenades us with his original songs while strumming on his guitar, Lyza's fave. When the last dish is stowed, once in a while the crew, minus Jack of course, manages to pull off a rousing game of Rummikub or Boggle.

At last the sea day is done, except for the one on watch. The sun slips over the edge of the world in a blaze of colour that deepens into jewel tones. The night sky becomes increasingly spectacular as we move further toward the tropics. The stars begin to appear one by one, as if an unseen hand is lighting our way with candles. Then, seemingly all of a sudden, you look up and see the heavenly dome afire with millions of them. The first watchman settles in at the helm for his three-hour stint, and the rest of us retire to our quarters to read or sleep. Our ghost ship *Porpoise* glides on through the fathomless chasm of the phosphorescent deep under the limitless sweep of the celestial spaces above. Tomorrow we do it all again, a repetitive but satisfying routine.

Cast Off Your Bow Lines

PPSS

"Believe me, my young friend, there is *nothing*–absolutely nothing–half so much worth doing as simply messing about in boats."
                  Water Rat (Ratty) to Mole in *Wind in the Willows*
                                    By Kenneth Grahame

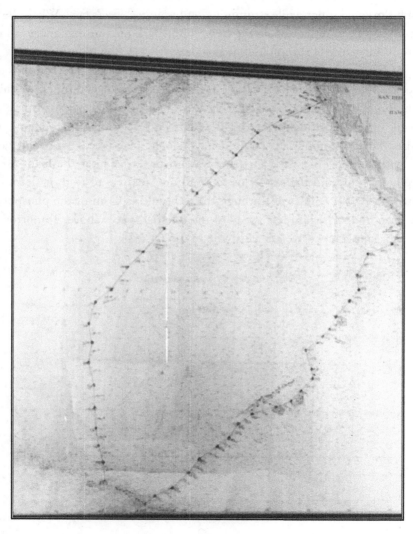

Our offshore voyage track—every dot a day
Outward bound: Neah Bay, Washington to Hilo, Hawaii
Homeward bound: Hanalei, Kauai to Sitka, Alaska

16. Hilo Heaven

July 10

I can hear you now, dear friends: "July 10! Your last entry was May 24. What happened to a month and a half?"

Okay, you have reason to wonder what happened to us. I'm feeling guilty and rather sheepish, but there are no excuses. What can I say? In a word, what happened to us can be summarized in one word—Hawaii! I know what you're probably thinking: *What could be so pressing in Paradise that we can't get even a little note, a scrap tossed off the tropical table to your faithful followers? Especially when here we are slogging away in our mundane existence while you turkeys sail the seven seas on Porpoise.*

All I can say is, "Sorry!" Please allow me to make some redress by giving a belated account of the last six weeks. I'm writing this with my feet propped on the rail, anchored in Hanelei Bay, Kauai, as *Porpoise* rocks slightly on the gentle swells. Beyond my brown bare feet, a sandy crescent beach fringed with palms and bungalows frames an impossibly picturesque scene of tropical blue seas, steep green hills, and scantily clad beach babes walking the shoreline and lazing in the sun. The more athletic and adventurous are surfing the swells crashing into the western edge of the bay, while, on the hill above them, the only major resort hotel has the monopoly on this premier spot.

Banished from memory are the huge scary seas and arduous living conditions it took to get here (how quickly we forget). We have exchanged that challenging passage-making for the vagabond gypsy life of island hopping and shore wanderings. As we meander each day through the Hawaiian archipelago, I wonder how we find ourselves living on *Porpoise* for free, with a cheap front row seat on some of the most coveted real estate on the planet. I guess the price of our ticket was not all that cheap, considering the cost of equipping the boat to

Cast Off Your Bow Lines

make the journey and paying our dues on a challenging ocean crossing. From my present vantage point, it has been absolutely worth it all.

Now that we're here, we've discovered a few downsides to sailing a recreational boat through the Islands. One of those is coming to terms with how inhospitable to boaters these islands really are. There are few marinas, and most of them have no space for transient vessels. Unlike our northwest coast, we find almost no protected places to anchor a boat, because most of the coast is exposed to open ocean swells, and many of the few possible coves or bays have anchoring restrictions aimed at curbing the damage of anchor chains on the coral. Nevertheless, we found a few places to cast the hook, even coming to rest in calm conditions for a time in front the famous Kaanapali hotel strip on Maui. At those times, we certainly considered ourselves part of the privileged set. Anyway, I'm getting ahead of myself. Let me first take you back to our making landfall at Hilo on the Big Island.

As we were told to expect, the trades held steady the rest of the way, making it a sleigh ride of a sail all the way to our destination. During those days, we often clocked over 150 miles every twenty-four hours, putting us ahead of schedule. On the last day before reaching landfall, I was on watch at 2:00 a.m. when a luminous glow appeared on the horizon ahead. At first I thought it must be my imagination, or that maybe I'd fallen asleep and was dreaming—a phenomenon that did happen during the long night runs when the line between consciousness and slumber became hazy. Or could I have fallen overboard, and the light I was seeing was the celestial city of my heavenly reward? Just kidding! But with no defined horizon in the dark, it appeared that the glow was suspended in the firmament like a heavenly city. I was glad when Chris came on watch and confirmed that I wasn't dead or hallucinating. I stayed up with him for a while, and soon we both could distinguish little pinpricks of twinkling light. The lights remained stationary with more appearing all the time. When I finally turned in, it was with the giddy conviction that we were closing in on Paradise. Excited as I was, I couldn't keep my eyes open and had no cognizance of Lyza rising for her early morning helm watch.

My first waking awareness was her voice, however, coming from the deck above my berth. As you've probably picked up by now, Lyza is one of those people who, when excited, has a way of infecting everyone around her with her enthusiasm. In this case, the infection had spread to an all out crew epidemic, and I didn't want to miss the moment. I jammed my legs into my Levis and stumbled

Hilo Heaven

on deck. Matt pointed toward a dark shape off the port bow whose contours were illuminated by a ghostly full moon. Land ho!

I was unprepared for the bombardment of my senses—especially the fragrances wafting on the air, like the most exotic, rare, complex perfume money could buy. I'll never forget that tantalizing aroma—filled with the sweet scent of plumeria and jasmine, and spicier notes I couldn't identify. I sucked in large drafts of the pervasive air and ended up with the same silly grin plastered on my face as the rest of the crew wore on theirs. But as overwhelming as the fragrances were, the sights vied for supremacy in this battle of the senses.

Just as I turned to look behind us to the east, the sun burst from the horizon as an orange orb, like a sunset in reverse, setting the water on fire. Swerving my head around to see the effects of the sudden burst of light on the land before us (that had been in shadows until this moment), I was gobsmacked. The face of the moon blanched and fled before this golden lord of the sky, who demonstrated his dominion by painting the shore ahead in almost unbearably vivid Technicolor. I felt like Dorothy stepping from black and white into the colour-filled landscape of the land of Oz. The green—the emerald green—I can only make a feeble attempt to describe.

Without realizing it, we'd become colour-starved for earthly tones after endless vistas of greys and blues of sea and sky, so suddenly we felt such gratitude for the sight of trees and grass, mountains and rock, even dirt. I was reminded of those images of people coming ashore after a long voyage and kneeling to kiss the ground under their feet. I got it! I groped my way forward to the bowsprit to get as close as I could to the sight of Mauna Loa, the massive volcano with its vast slanted slopes, looming larger and larger.

Jack, bless his heart, had told me that the first time he sailed to Hawaii, he'd been sorry the ocean crossing had ended—he had just wanted to keep on sailing. As you know by now, Jack's a piece of work. At that point, I considered his preference inconceivable as I stared at my deliverance with rapt beatific countenance, as though I'd finally entered the realm of just men made perfect. I was only aroused from my stupor when Lyza came forward to confer on me the captain's honour of bringing us into port, like Joab setting David at the head of his armies to take the city.

We took down our sails and started the motor in preparation for mooring. As we rounded the breakwater, the seas instantly became a flat calm. We had sailed on rolling water for twenty-three days, and we had made it to our destination. And in a pretty respectable amount of time too. I noted the date for the log

Cast Off Your Bow Lines

book—May 31. We found our way into Radio Bay and tied up Tahiti-style, our bow anchored out and stern secured to a concrete pier. There were only a handful of other pleasure craft lined up.

The rest of the crew thought this mostly commercial little port was a bit tacky, but I loved it and revelled in the ability to step off the boat onto dry land. Before the rest of the crew could disembark, I walked to the government office to clear us in and assure the authorities that we weren't carrying any invasive species of flora or fauna to this isolated environment. We were the only non-native species about to invade, and we immediately rented a car to fulfill the many reconnaissance missions we had planned for our time on the Big Island.

The long passage was over, and suddenly our shipboard routines were exchanged for the hustle and bustle of landlubber life. Within a couple of days, we managed to get Jack off *Porpoise* and on a plane back to the mainland, and our daughter, Julie (Chris's wife), and their six-month-old baby, Kai, off a plane from the mainland and onto *Porpoise*. A friend, Jocelyn, also came with Julie to help on the plane with Kai, and her reward was to get to tour the Big Island with us.

Now that we were on shore, we lavished ourselves with hot showers in the port's restrooms and filled the boat's water tanks to overflowing. We were genuinely proud of our rationing efforts—we still had 150 litres of water left at the end of the voyage, and we had started out with a thousand-litre supply. After three weeks of miserly water allowances, we became profligate, splashing water all over like sparrows in a birdbath, rinsing the salt from our skins and the decks.

When we checked our diesel tanks, we discovered we had only used 193 gallons traversing 2,600 miles of ocean. It was amazing to us that we were conveyed here almost entirely by the power of the wind. That feat may have been accomplished for thousands of years, but it was a marvel to have done it ourselves.

Our forays into the Big Island lasted only four days, ranging from the depths of the Pololu Valley, where an entire village was destroyed by a tsunami in 1946, to the heights of the observatory on Mauna Loa. Did you know that Mauna Loa is higher than Mount Everest if you measure from the bottom of the ocean to its top? This isn't a travelogue, so I won't go into all the details, but suffice it to say it was all fun, fast, and foody.

All too soon it was time to shift *Porpoise* out of the cozy harbour and head around the north end of the island to Kona on the leeward side. We embarked at 2:00 a.m. in order to round Upolu Point, the northernmost tip of the island, in daylight. It was a rude baptism for Julie and Jocelyn as we left the protection of

Hilo Heaven

the breakwater and had the trade winds smack us in the face. Kai, however, was unfazed by the turbulent seas (more on our little sea baby later).

It was rough and rainy with poor visibility for four or five hours. Julie and Jocelyn were soon on deck making their own belated offerings to our patron saint, Juan de Puca. They eventually collapsed on the cushions of the back bench in the tropical downpour, with their faces pea green and their long wet hair hanging in bedraggled strands around their heads. When we were finally able to ease the ship broadside to the wind, our speed increased, but the large beam seas still left *Porpoise* heaving to and fro (along with Julie and Jocelyn). Finally rounding the northern end, our rollicking movement smoothed out into a lovely downwind jaunt in the spritely breeze. As the sun rose to warm our chilled bones, both the girls' spirits rose too, and their stomachs settled.

The incredibly lush coastline slid past on our port side, with steep volcanic valleys penetrating deep into the island's mystic interior. Numerous waterfalls snaked down the hillsides and cascaded over cliffs, sometimes plunging directly into the sea. Spirits were high as we progressed toward Upolo Point and soared as we rounded it, racing smoothly along beside the drier sights of the lee side of the island. The day was a long twenty-hour run before we found our way as darkness fell into a little commercial breakwater, where we sought refuge and calm anchoring for the night.

The next day we ran down the coast to Kona. There we were delighted when, in the midst of setting the hook, a small dolphin burst completely out of the water and twirled around before diving back underwater. Spinner dolphins! At first sight of them somersaulting twenty feet off the boat, Chris went AWOL, tearing off his tee-shirt, grabbing his snorkel, and leaping overboard. He was determined not to miss his lifelong dream of swimming with dolphins. Lyza, always a porpoise and whale nut, was soon right beside him, squealing through her snorkel as the spinners skimmed along under her, close enough to touch. For the next week, those dolphins were the main attraction with their daily visitations, sometimes as many as a couple dozen at a time. It just never got old watching them burst out of the water, spin around several times, sometimes in combination with a somersault, and splash back down. I suspected they were on the resort payroll the way they snuggled up to any tourist with a snorkel, but the locals assured me these visitations right in front of their town were rare.

We rested at anchor for a whole week, enjoying the town, swimming in the warm water, and going out to eat. We also shopped for groceries and more groceries and more groceries. I'm not sure where it all went. Matt and Chris stay

Cast Off Your Bow Lines

as lean and toned as ever. They think they're pretty hot stuff with their tans, sun-bleached long hair, and sunglasses, and Julie and Jocelyn belong on the cover of *Beach Babes of Kona* magazine. It was with great reluctance that Jocelyn flew back to Vancouver at the end of a glorious week of fun in the sun.

She had to fly home to experience trials and tribulations with our cat on the home front. The cat had to be sprung from our house because the June renters had an aversion to felines. Miyagi was quite offended, as he considers the house to be his, and found himself shipped off to Jocelyn's apartment when she got home, where he registered his displeasure on her best couch shortly after his arrival. It was then discovered that he had imported fleas. We felt terrible about our disrespectful cat. We think Jocelyn may have second thoughts about mixing with the Clarkes and their hairbrained adventures (a conclusion perhaps some of the rest of you have come to).

After she left, we made preparation to cross over to Maui via the infamous Alenuihaha Channel—hard to pronounce and, by reputation, hard to cross. It's even sometimes referred to as the most dangerous channel crossing in the world. (We sincerely hope this is hyperbole.) It's only twenty-six miles wide, but the huge volcanoes on either side of the channel funnel and amp up the trade winds, causing very rough conditions most of the time. The name itself created in us a suspicion about that channel getting the last laugh on sailors daring to cross it. You crossed the whole ocean only to be knocked down here between the islands. Ha ha!

Following the marine forecast closely, we chose what sounded like a mild weather window early in the morning to dash across that trepidation-producing chunk of water. The forecasted winds of twenty to twenty-five knots picked us up like a kite to skip over the robust waves. By mid-channel, the winds increased to thirty-five to forty knots, contrary to the mild forecast.

The crossing seemed over in a short time, as we fairly roared across under just the staysail and mizzen, all of us holding on tight and doused by the warm spray over the deck and into our faces. Rather than fear-producing, it was exhilarating. Speaking of exhilarating, the more the wind picked up, the more our littlest sailor expressed his elation. Grandson Kai is quickly developing his sailing chops and easily adapting to life at sea. Not only does he seem completely comfortable with the rolling motion of the boat, but he has added the acrobatic feat of leaping about in his jolly jumper in sync with the heaving waves. Chris rigged up the jumper in the salon, and Kai bounces crazily forward and back while singing at the top of his voice. During our wild rides across a number of boisterous channels between the islands, he loved to be outside on deck as he balanced his muscular

Hilo Heaven

little body with one hand tightly clamped to the rail, shouting joyfully the whole way. (Don't worry—we make sure somebody always has as firm a grip on him as he has on the rail.) He maintained this pose for long durations, the brim of his sunhat flapping up and down over his eyes and a gleeful grin plastered on his face.

With Kai keeping us all entertained with his antics, we crossed the Haha Channel, as we came to call it, in short order, and we soon found ourselves near Maui, with the waves crashing hard on the southern shore, spouting up in great geysers against the sharp black lava rocks. The further we went, the more we felt the effect of being on Maui's leeward side, and we finally had to resort to motoring the last mile to the crater of Molokini, arriving in the late afternoon just as the last tour boat loaded with snorkelers was leaving.

The boys needed no persuasion to dive into the aquarium-clear water and fetch the tour boat tether to underwater moorings. Large purplish fish that clustered curiously about them as they worked at their task intensified the feeling of being in a giant aquarium. The creatures who call this place home are so conditioned to people that they didn't even see us as an intrusion.

As evening settled in, it was amazing to have this popular and unique enclave all to ourselves. It was a far cry from the first time Lyza and I came to Maui on our twenty-fifth wedding anniversary. Back then, over twenty years ago, we got sucked in to one of those ubiquitous timeshare presentations, but the upside was that one of the offered enticements was a snorkel trip to Molokini Crater. That sounded exotic to us, but it was definitely made less so by the many other tourist boats and snorkeling bodies we had to share it with. The only others now sharing our idyllic spot on this magical evening were the boobies, shearwaters, and frigate birds who nested in the cliffs and soared over the volcanic crescent they claimed as their own, as well as the myriad of marine life under the glassy water.

In the distance, the familiar contours of Maui beckoned us. Lyza and I know Maui well from the many visits we've made after that initial anniversary trip. It has become one of our favourite places on earth, but every other time we had arrived the normal way, disembarking from a jet and exploring the unique topography by car. Approaching by boat gave us a whole new appreciation of Maui's many charms. It was easy to relate to how those early whalers must have felt landing here after being at sea for months or even years. Lahaina Town, where we were heading tomorrow, became a magnet for those wandering sailors, who became intoxicated by this paradise on earth.

Probably those mariners felt a little like we do now, transported to this tropical Eden aboard our own Pacific Northwest sailing ship. It's like we're aliens

147

Cast Off Your Bow Lines

visiting another planet, or like the ants we've noticed lately on the boat. These critters are aliens on our Planet *Porpoise*. They've been stowaways since our stern tie at Hilo Harbor, where I discovered them boarding us by way of the yellow shore power cord (that turned just about black with their scurrying little bodies). These pirates hustled across their gangplank to raid our stores and carry off whatever booty they could back to their lair. We tried to put a stop to this thievery by blocking them from the cord, but it was too late. They had scattered below decks so were now captives cut off from any way of escape. Once we left Hilo, I figured these foreigners would eventually die a lonely death in the bilge, being cut off from their queen, albeit comforted by the largesse of our larder. (As I write this on July 10, several weeks later, I am still squashing the little blighters every time I see one, which is often. I'm praying they didn't smuggle a new little baby queenie of their own on board to set up a new colony on the island of *Porpoise*. It's no wonder there's such concern about invasive species coming to these islands. However, in this case, the native species has invaded us.)

The next day, we sailed into the Lahaina Roads—not the asphalt kind, but "roads" is what you call open water where boats can anchor. We needed a safe place to moor *Porpoise* for a couple of weeks because the rest of our family, including all kids and grandkids, were flying into Maui for a family rendezvous. I had an idea to try regarding moorage, so I started up the motor on our inflatable.

I pulled the dinghy up on to the sand, approached the rather rustic building that housed the Lahaina Yacht Club, and walked up to the lady manning the desk. Wishing her a good morning, I pointed out the window to our lovely little ship and said, "I've come in from that ketch out there to see if there are any buoys available."

"Are you a yacht club member?" she inquired.

I was ready for this question. "Oh yes, we belong to a great little club in White Rock, British Columbia." I pulled out my card, but I doubt she recognized the name as a province in Canada (many Americans don't, I'm afraid), but it still sounded good to the ear. Anything with British in it sounds a bit exclusive, doesn't it? What she didn't know was that the White Rock Yacht Club was the most one-horse yacht club on earth, having no clubhouse, no marina, and practically no yachts. What it did have was a very optimistic leader named Tom who didn't seem fazed by our club's lack of substance and kept us going year by year with his sheer enthusiasm. Now my faith in his leadership paid off as I was able to flash the shiny card he'd issued me years previous. I hoped she wouldn't notice it was expired.

Hilo Heaven

She smiled at me brightly, taking in my neatly trimmed white beard and Greek fisherman's hat. I've often been asked if I'm Captain Highliner, the mascot of the frozen seafood company of the same name. I decided to work that angle and attempted a bit of seaman's talk with a touch of an Irish accent.

"Aye, we've just hove in from a long trick across the great wide ocean, and what a wonder it is to arrive upon your lovely shores. You're a sight for sore eyes to be sure. Would ye be honouring reciprocal privileges to yacht club members from foreign parts?" I probably did more harm than good with that little sally, but she remained friendly.

"Uh ... sure. Yeah, we do that ... don't we, Fred?"

The affable male behind her shrugged his shoulders and nodded at the same time. I decided to bolster my case with a little namedropping to seal the deal.

"Do you know Sandee Butterley? She's our good friend and has a yacht club membership here. She suggested we check in with you when we got to Lahaina."

Sandee is the type who is on good terms with everybody, and even though she didn't own a boat, she was a frequent customer in the club's for-members-only dining room. Clearly, I had used the magic word when I mentioned her because in very short order we had a complimentary two-week berth on a buoy in front of the yacht club. The yacht club proved to be a strategic location in the middle of the town's busy waterfront, and when shopping and playing ashore, we could get out of the sun by popping into its cool, shaded interior for a beer while keeping our eye on *Porpoise* right out front.

Sandee made her condo available to us, located a few miles up the road, and the offer couldn't have been more appreciated. For the first time in many weeks, we had our feet planted on terra firma, and we revelled in condo living at Hoyochi Nikko. Lyza and I took the first stint, and Chris and Julie the second. The others joined us for barbeque dinners at the big table by the pool and swims at the beach in front. The time flew by as we packed this special time with outings and explorations. A particular highlight was when the whole family went snorkeling with the sea turtles in a little bay nearby. It was pretty thrilling to have these magnificent creatures swim up and look you right in the eye as they gracefully slid past only a foot away.

All too soon the crew had to say goodbye to rest of the family, and our bow pointed away from Maui toward Molokai, an island we saw every day from our condo view. Finding tenuous holding ground in the shallow waters behind a small breakwater, we walked the long dock into the tiny town. Molokai has a

149

Cast Off Your Bow Lines

much slower pace of life than Maui and hasn't become such a big tourist attraction, but it has its own quaint charm and really good ice cream.

After an overnight stopover, we were on our way to Oahu. We tackled another notorious waterway between Molokai and Oahu, ripping along under only staysail and mizzen in a forty-knot wind. Curiously, this was the only channel in all our island hopping where we saw another sailboat under sail, and it was a reassuring sight. We hoped we looked as stunning as they did surfing along with white sails flying.

By early afternoon, we flew past the famous sights of Diamond Head and Waikiki Beach, and the many hotels that fronted the beaches of Honolulu, turning into Ala Wai Boat Harbor. In spite of its modern city setting, the marina's facilities for transient boats were alarmingly primitive. We were directed to tie up at a derelict pier, in the Tahiti style, stern to, and we literally had to walk the plank to get on shore (and a narrow one at that). One positive was the interesting assortment of fellow seafarers lined up beside us. Our closest neighbour, Spike, had just sailed in on his catamaran after five years of cruising in the South Seas and was chock-full of sea stories.

As a native Hawaiian, he helped us find our way around Honolulu, giving us directions to the grocery store, the mall, and even a chandlery for the spare parts we needed. I'm afraid it was quite a culture shock for us as we headed on foot out of the marina, under the archway of the Hilton Hotel, and directly into the middle of a full-blown concrete jungle, complete with traffic noise and fumes. It was a jarring contrast from our vagrant gypsy, sea-going life. Even so, we enjoyed our brief foray into this metropolitan centre, loading up on tropical fruits and goodies and boat parts. Julie even checked out the nearby shopping mall and came back with some great deals from J Crew.

Despite its deficiencies, one advantage of our moorage was its close proximity to a lovely waterfront park, where we took Kai for walks and picnics on the grass. It was also the site of an amazing fireworks display on the evening of the Fourth of July, our designated departure date for our next hop. It didn't take much imagination to figure the city was bidding us a fond farewell with its best display of rockets, starbursts, and shimmering colour-changing waterfalls. There were even fireworks that turned into glowing happy faces in the sky, a sight we had never seen before. It wasn't hard to pick our way out of the little boat basin lit by the pyrotechnic explosions. The date of this celebration reminded me that we had only three more weeks to go before our exodus from these wonderful

Hilo Heaven

islands. I felt a nervous little jolt as I thought of the intervening swells we had to face before once more touching our native land.

Night runs were no longer intimidating after our many weeks at sea, and we watched for other vessels as we motored through the local shipping lanes, eerily backlit by the Honolulu skyline. By morning, we sailed under bright blue skies, Oahu a fading silhouette behind us, and Kauai invisible under a clot of cumulus clouds about a hundred miles ahead of us. By afternoon, we could make out some volcanic shapes, and by dinnertime, the brisk trades had brought us alongside the verdant green coast of the island some consider the true gem in Hawaii's crown. Before dark, we entered Hanalei Bay on the north side, anchoring in thirty feet on a sandy bottom clearly visible in the crystal water.

This was the most visually stunning bay we had seen yet, the islands saving the best for last. The turquoise waters are encircled by a crescent-shaped white sand beach surrounded by soaring jagged peaks whose tops are hidden in the clouds. Waterfalls fall in the distant valleys and run down through a great variety of tropical greenery until the splashing streams flow out at the foot of palm trees into the bay. Development is modest, with the one large resort and a number of bungalows built around the perimeter of the strand. The fragrances waft to our anchorage and are even more aromatic than at Hilo. We've been stationed here now for the last week, except for one ill-advised morning sailing excursion down the Na Pali Coast (where they made the *Jurassic Park* movies). It was an easy sail downwind for an hour, but the return trip was a four-hour bash upwind. For a while it was nip and tuck whether we'd be able to claw our way back. Sailing the trades is generally a one-way trip.

Here's a little interesting side note—Matt has become a believer in miracles after flipping his kayak over and losing his expensive Maui Jim sunglasses. The glasses were perched in his waving mane of hair when it flipped, so they sank to the bottom, down amid the sand and coral. Chris and he spent the next hour diving around in the water trying to find them, but no luck. Not one to give up easily on such an iconic piece of his beach bum wardrobe, Matt got up early the next morning and paddled off for another go at finding them. The odds were greatly stacked against him, since the stream mouth where he'd lost them was the only murky water in the place. Nevertheless, on his first dive down with mask and snorkel in the approximate location of the mishap, the first thing he laid eyes on? His sunglasses perched on a ledge of coral! I'm not sure if he claims to have prayed or not, but regardless, it would appear God not only knows when a hair

151

Cast Off Your Bow Lines

falls from your head, but He also takes note of when and where your articles of clothing fall, or in this case, a cool pair of polarized shades.

Hanalei Bay for us has been nothing less than a quintessential paradise. None of us want to leave, although we know we have to soon. The other day I caught Julie and Chris on the back deck plotting to sell all and move here. I don't think it's likely, but I have to admit we've all been tempted with such fantasies. The reality is that tomorrow we rent a car to drive Julie and Kai to the airport to say goodbye. Then we need to find the Costco on the citified other side of the island.

Over our shoulders to the north, we hear the roar of breakers on the coral reef, and beyond that the vast blue endless Pacific awaits us. In a few days, we'll turn our backs on this nirvana and point our bow into that void, and there are a few whispering uncertainties in the back of our minds. Lyza is hoping the westerlies will cooperate with us to get us back home comfortably. (Jack had some miserable stories of some of his passages back). Chris is hoping he won't experience that horrible seasickness once we hit the open ocean again. I'm assessing our prowess at crossing the ocean without our mascot, Jack. Strangely, I don't feel as apprehensive as I might. I actually feel some excitement and a sense of rising to the challenge. Before leaving the mainland, Lyza spent hours poring over *World Cruising Routes* by Jimmy Cornell. It became part of her dream and master plan for us to take the great circle route back, landing in Alaska. She loved the romance of sailing through the Hawaiian Islands and sailing right up to a glacier in Alaska, all in the same summer. The prospect of seeing glaciers was interesting to me too, but I was particularly looking forward to another run down the Inside Passage once we reached the coast.

So you may not hear from us until we make landfall in Sitka, Alaska, approximately three weeks from now, but you can follow our progress on our Spot GPS tracker. I'll send this report off while we're still in signal range.

Hilo Heaven

PPSS

With favoring winds, o'er sunlit seas,
We sailed for the Hesperides,
The land where golden apples grow;
But that, ah! That was long ago.
How far, since then, the ocean streams
Have swept us from that land of dreams,
That land of fiction and of truth,
The lost Atlantis of our youth!
Whither, ah, whither? Are not these
The tempest-haunted Orcades,
Where sea-gulls scream, and breakers roar,
And wreck and sea-weed line the shore?
Ultima Thule! Utmost Isle!
Here in thy harbors for a while
We lower our sails; a while we rest
From the unending, endless quest.

Excerpt from "Ultima Thule"
By Henry Wadsworth Longfellow

The whole family meeting up in Maui at Sandee's condo

17. Captain Bluewater

July 21

From my previous musings on the cusp of setting out from Kauai, you may wonder how Captain Highliner is faring on this homeward track, especially without his old faithful, Smilin' Jack. Well, I'm here to testify that miracles still happen. I'm not sure why, but everything is different. It happened when I closed myself into the closet-sized head as we moved out of Hanalei Bay and into the unpleasantly familiar rollers. Behind me were the palm trees, the gorgeous sand beach, and the quiet anchorage. Ahead of me was nothing but 2,700 miles of open sea. I locked the door and was about to have a private breakdown right there when, suddenly, I stopped short of this negative bent, braced up to my realities, and gave myself a good talking-to.

The pep talk went something like this: "This is what I'm doing right now … what I'm meant to be doing … what I chose to do. I can do this! Of course there will be challenges and difficulties—we're crossing an ocean, for heaven's sake, but I'm going to suck the marrow out of every single day of this experience. The ocean is going to be my fundamentally good-natured rollicking friend, and we're going to get along, even be good buddies. That's how this is going to go, okay?"

I stepped into the head as mild-mannered Cap'n John and came out of it none other than Captain Bluewater, sailor of the seven seas (well, one, anyway). It's been eight days since my telephone booth experience, and I must tell you that so far I've been thoroughly enjoying myself. The ocean is the same—chaotic as ever, blue and wet, knocking us about in our little shell. But I am different. I'm actually revelling in this leg of our journey. Oh, I'd still prefer a nice anchorage every night, but this kind of sailing has its own unique allure. One has to listen to the ocean's music and learn to move to its rhythms.

Captain Bluewater

I'm putting on my thinking cap to figure out what caused that attitude-altering change, because it might be applied to lots of life situations, I'm thinkin'. Maybe the spirit of Jack has come upon me. I haven't yet said, as he did regularly, "I just love it out here," but I have to admit that I actually do. It's possible that when we get to Alaska, I'll be sorry the passage is over. I've taken to sitting for hours at the helm with Jack's silly grin on my face, just gazing at wave after wave and marvelling at how *Porpoise* slides and glides over them. I hang on and exult in the ride. When she heels over in a gust of wind, I hunker forward like a jockey in his stirrups, giving my thoroughbred her head and letting her find her own stride. We gallop along at breakneck speed, hour after hour after hour. The other day they had to drag me off the helm just to come in for dinner.

Mal de mer? Scrambled brain? A thing of the past. About the spirit of Jack coming upon me—don't worry, I haven't taken up his happy hour rum and coke habit (yet), but I do start getting a little nappish at two, fiddly at three, and talkative at four.

So far the conditions have been good and we haven't met any serious weather, so my grand attitudes could deteriorate rapidly should the ocean become unkind. I also have to admit that I like being on the homeward stretch to my beloved Pacific coast to the north. My true home is among the myriad of fir-clad islands and sheltered inland waterways for sure. But for the present, I'm content to savour each day of this exposed ocean life. Carpe diem!

I'm trying not to be smug about my newfound buoyancy as I go whistling about the boat. I sometimes catch Lyza looking at me a little puzzled, wondering what happened to the grouch. Captain Bluewater is a welcome replacement, but a little bit annoying. I swing like Tarzan through the galley, one handhold to another, landing nimbly behind her with a grin and a nuzzle of her neck. I check to see if she needs my help with anything before dashing topsides to check the sail trim. I'm not taking my newfound seaman's grace for granted, but it sure is nice to feel really good in my present circumstances.

I've decided the ocean, although often unmannerly and sometimes scary, is a compatible enough companion if you just accept her as she is (possibly also good advice in domestic life as well). I've heard and read enough about her cantankerous side to know I would rather not live through it firsthand. It's enough to read other people's ocean war stories—I can manage fine without any Jonah experiences in my repertoire of sermon illustrations, thank you very much. I hope to be able to report an uneventful, even boring, ocean crossing when we get to Sitka.

Cast Off Your Bow Lines

The crew is compatible enough. Lyza, having grown up with three brothers and no sisters, likes masculine companionship, especially that of her boys—Matt, Chris, and me, of course. Both Chris and Matt are laid back types and easy to please, as well as being very willing workers, so a person couldn't ask for a better crew.

We're definitely missing Jack, but on the other hand, I kind of like being the undisputed captain. Without Jack around, I make all the calls on sail changes, observations about weather, where we should head, and how we should catch fish. I've even taken to grimacing and stroking my chin when someone asks me a question before giving them Jack's trademark, "Waaaalllll ... now that's a tough one." Then I laugh in his "nobody knows for sure" sort of way.

The boys humour me, but it's obvious I'm not quite as entertaining nor have nearly the same stock of war stories to tell. We imagine our guardian angel Oddbody is still watching over us, and we still send out our coordinates through the sat phone and wait for weather info to come back from Jack's friend Al. Jack took off on vacation just as we left Hawaii, so we haven't heard a thing from him, but we already know what he would say: "Well, you pretty well gotta just take what you get anyway. There's nothing you can do about it. Good thing she's a lucky ship."

So much for worrying about weather reports.

We settle into a contented routine of simple existence divorced from the insistent urge to always be doing something. You can't be anxious to get anywhere or meet schedules and deadlines, because *Porpoise* ain't going to go any faster, nor will the ocean get any smaller. Each moment is everything and nothing; our breathing, like the rising and falling of the surrounding seas, defines the slow rhythm of living in the moment. Time ceases to tick like a clock and becomes fused into the endless rows of waves behind and before. It has a way of becoming simultaneously shortened and lengthened out here, kind of like what I think eternity will be like. It seems we have always been here and will always inhabit this space. It's hard to explain.

The shifts on the helm, watches of the night, the meals (especially the meals), the reading of books, and the slumbers at all hours determine the cadence of our days and nights until we wonder if this is all there is and all that ever will be. Nights are a wonder under the Milky Way on a moonless night, when no artificial light dims the brightness of the heavens. Under sail there are no sounds but the wind in the rigging, the rush of water on the hull, and *Porpoise's* arthritic

Captain Bluewater

creaks and groans. You are alone, and your aloneness is a good place where God waits and watches.

When daylight touches the horizon, it's at first so subtle you think you're imagining it. Then wave tops become discernable as the light collects in a gathering place on the sea in the east. That light changes colour from pale to yellow-gold to orange to red, and then the psalm comes to life: "*In them hath he set a tabernacle for the sun, which is as a bridegroom coming out of his chamber, and rejoiceth as a strong man to run a race*" (Psalm 19:4b–5, KJV). Another day has begun. The night watchman shakes the cold from his limbs and removes his hood so the morning rays can warm his body and soul. Then he hears the first stirrings of the rest of the crew and the words that are music to his ears: "Coffee time!"

Along the way, the sea has yielded up its bounty to us, and we landed three dorados (also called mahi-mahi) while still in tropical waters. We caught no tuna this direction, but much to our surprise, we brought in a salmon on our tropical lure, called a Rapala plug, five hundred miles off the coast. What royal feasting it is to furnish fish from the sea by your own hand as we mosey along! We can't get over the dorados with their intense electric blue and green colours as we reel them in close to the boat. As we scoop them up on deck, their iridescent brilliance is on full display but quickly begins to change out of the water. They go through several hues until fading to a dull yellow-grey in death. It's sad to watch and seems a violation to catch these gorgeous fish, but we overcome our qualms after tasting their succulent flesh. (All except Matt, who then and there became an avowed vegetarian.)

Speaking of Matt, he keeps insisting we aren't drilled enough in safety procedures, especially in the matter of saving a man gone overboard. Our usual volunteer for rescue drills is George, a floating man-overboard pole with an attached red and yellow distress flag. We deploy him regularly on sail training charters to teach rescue techniques to would-be sailors. One day we hove to in order to land one of the mahi-mahi, Matt decided to take a bucket bath for some reason. This is not unusual behaviour for him—he's always dunking or splashing himself whenever it strikes his fancy. While scooping up a bucket of sea water this time, the carabiner on the bucket lanyard malfunctioned, and the pail slipped its line, plopping back in the water, and floated immediately out of reach off our stern.

Thankfully, Matt didn't follow what might have been his first instinct and dive in after it. But he did see it as a great opportunity for practising a man overboard drill. I would probably have just let the stupid bucket go, but who am I to discourage good safety practices? So we swung the boat around to rescue the

157

Cast Off Your Bow Lines

bucket in a twenty-knot wind. We made a cursory pass by the pail and missed it. And then in the bouncing waves, we lost sight of it altogether. Now it's out there with Wilson floating on the seas. This unfortunate conclusion to the matter only confirmed to Matt that if he ever fell overboard out here, he would likely suffer the same fate. Jack's pessimistic words on the subject came back to haunt me: "If you fall overboard, just sink fast, 'cause nobody's gonna be able to come back and get ya."

The transit this direction can be notoriously nasty just about any time of year if successive Aleutian lows decide to roll over the top of the North Pacific High, but so far, we've lucked out with the trade winds serving us well. From Kauai we made a heading almost due north, trying to stay three or four hundred miles from the centre of the North Pacific High, where the winds can be capricious or non-existent. Where we are the trades blow steadily and, riding off the edge of the High, our strategy is to keep out of reach of the periodic lows that could ruin our peaceful run.

Often the first leg of this homeward passage can also be a windward bash into the northeast trades, especially if they're blowing hard. Thankfully, they blew gently for us at first, a soft fifteen knots more on the beam than is usual, so we were able to hold a more northerly heading. Sometimes you have to almost head for China at first and then double back east once you hit the prevailing westerlies above the High. We only gave up two degrees longitude (120 nautical miles) over ten days, and we quickly made up the lost miles once we were far enough north to make the dogleg to the east. Every day the water and air got a few degrees cooler until, reluctantly, at what we hope is about our halfway point, we packed away our shorts and sandals and hauled out our socks and parkas.

Once we turned east toward the coast, we sailed into thick fog that lasted for days, taking the edge off our idyllic crossing. At times visibility dropped to zero and the air got much colder and damper. We began to feel as if we'd drifted into some eerie twilight zone from which we would never return. For quite a while the wind kept up, a phenomenon that almost never accompanies coastal fog, and we just ghosted along hour after hour. We were deeply grateful for our radar and AIS technologies, completely trusting that those little red triangles on our GPS would warn us of any approaching ships. What choice did we have? But, honestly, we were probably most grateful for a particular comfort feature more than the safety equipment that could save our lives. Our faithful diesel stove/heater in the salon took the chill off down below and became the warm beating heart of the boat in that penetrating damp.

Captain Bluewater

This fog wasn't a very cheery prospect, but it also wasn't the biggest challenge to my Captain Bluewater persona. That threat came from a different trial a few days after the fog settled in. The wind gradually dropped until we finally had to fire up the motor to get anywhere. We were rumbling along under engine power in the middle of the night when, suddenly, the high-pitched oil pressure alarm went off on the dash, red lights flashing ominously.

I quickly turned the motor off and got down into the engine compartment to investigate. Checking the dipstick, I was dismayed to discover that the oil didn't register on it at all. I unscrewed the cap, poured a couple of litres of oil into the engine, and checked the stick again. Still didn't register any oil. Now I began to feel a rising panic in my throat. Only a couple weeks earlier, at anchor, I'd checked the oil and marvelled at how little oil our trusty Ford Lehman 120 burned. It was a very reliable engine. Yet here we were, low on oil after only a few hours of running. What had happened to it? There was no telltale evidence of oil in the bilge to indicate a leak, and no blue smoke in our exhaust to indicate major engine damage. The missing oil had to have gone somewhere. Two gallons of oil later, I finally got a register on the dipstick, but that left us with only six more litres of oil on board. I was about to reach for the key to turn the engine over and get on our way when common sense kicked in. If I just turned the engine back on without getting to the bottom of this strange oil disappearance, it could happen again ... only next time we wouldn't have any oil reserves to replace it. At that moment we were in complete calm, a good thing for engine repair but a bad thing for sailing. If the doldrums we were in continued, we'd be dead in the water for who knows how long.

Standing there in the engine compartment, rolling about forlornly in the ocean swells, I sifted through my limited knowledge of the engine's mechanical systems. I knew something about the engine cooling system and came up with a hypothesis. There was a component on the side of the engine that ran salt water around copper tubes inside a cylinder to cool the engine oil. If the copper tubes broke down through electrolysis or wear, the engine oil would leech out and be carried in the salt water through the exhaust pipe into the sea. The oil could have been slowly seeping out over a few hours running and we wouldn't know it until the alarm went off. Thank God for the alarm system, or the engine might have finally overheated and seized up altogether. I put that thought out of my mind. I needed all my powers of concentration to solve this dilemma.

If my diagnosis was correct, I knew what to do. We'd just have to swap out the oil cooler, a task I'd done before. *Porpoise*'s previous owner mentored me in

159

Cast Off Your Bow Lines

proactive systems maintenance, admonishing me to always carry as many key spare parts as the boat could hold. It would not do to be broken down on a charter in the Canadian coastal wilderness or in the middle of the ocean with no spare part on hand. So I pulled out my spare exchanger, gathered my tools, crossed my fingers, and went to work. I just hoped to God my diagnosis was the right one.

Chris was on watch by then, but since we weren't going anywhere, he became my assistant—and a very useful one at that as we wrestled with rusty bolts and stubborn clamps, detached water lines, removed the old exchanger, and replaced it with a brand new one. I confess to having a moment of pride in knowing how to do the job and having been prepared for such an emergency. The only way to know if our fix was right would be to turn the ignition key and run for a period of time to see if the oil level stayed high. Just as we finished our job, a gentle breeze sprung up. I felt reluctant to test our fix in the dark of night, so I asked Chris to put up the sails, even though we only made three knots. Trying out the new exchanger could wait for the freshness of morning light. I needed to take a rest.

You're probably thinking, *Sounds like you came through that ordeal with flying colours.* If you thought that, you'd be wrong. You may think me a fairly brave fellow to be out here on the ocean in the first place, but there is a cowardly side lurking under the surface. Waking in that darkest hour just before dawn, I unravelled. When thinking about dire eventualities in the dark bowels of a boat, with the wind moaning in the rigging, dark musings can become desperate. I don't know what it is about that time between four-thirty and five-thirty in the morning, but I have come to call it the witching hour, the time when all the vulnerabilities of life become amplified. Let's just say that lonely hour has not usually been my finest hour.

The anxieties I felt that particular witching hour were perhaps nurtured by some recurring dreams that had haunted me most of my adult life. One was centred around the demise of our old sailboat, *Fred Free*, although oddly enough I never had that kind of dream about *Porpoise*. In the dream, I'd lift up the floor boards to discover water rising in the bilges, slowly engulfing the engine as I frantically bailed and bailed. Realizing I was losing the battle, I'd desperately search for the source of the problem and too late discover that teredo worms had eaten through the thick yellow cedar planks. The dream invariably ended with me treading water, holding a line still attached to my sinking ship that was now nothing more than a rotting skeleton. As my beloved *Fred* slipped under the waves, I could not let him go. I too would go down with the ship to my watery doom.

160

Captain Bluewater

The other recurring dream was about my vocation. I repeatedly dreamed of standing in our church building, preparing to speak, and looking up to discover that all but a handful of people had deserted me. In both recurrent dreams, the scene was accompanied with an ominous feeling that the end of the world had come, a malaise that would linger even after waking.

It wouldn't be hard to do a little psychoanalysis on me based on these dreams. Go ahead, if you're so inclined, but I'm just trying to give you a little background so you can understand why this incident affected me the way it did.

By morning, Captain Bluewater had been demoted to Captain Coward. My worst imaginings went something like this: What if my diagnosis of the problem was wrong? I had bought a new kind of engine oil in Honolulu, and the cashier had assured me it was right for my motor, but what if it wasn't and it damaged the engine? If the engine was damaged, then we were left to rely only on the power of the wind and would be subject to its vagaries. We could be stuck for weeks in the doldrums of the North Pacific High. Furthermore, if we lost the engine, we couldn't charge the batteries. We'd soon lose all the food in the fridge and freezer, not to mention that all our electronic gear would fail, including the GPS. (Lyza and Chris have been taking sun sights with the sextant, but that's been just a little, shall we say, unreliable.) We could be reduced to heading for the sunrise in the east with our compass in the hope of eventually reaching land ... somewhere. After spending that witching hour with these dire imaginings, I had us lost, drifting in the trackless ocean, starving and emaciated under a merciless sun.

I tried to shake off these dark forebodings, realizing I was getting carried away with my imagined scenario, but my bleak broodings only slid off in another direction—actually a more realistic one. Even if we managed to sail our crippled ship back to civilization, we would have to make for a place to repair the motor, so our Alaska sojourn was out of the question. It could take weeks to rebuild the engine, because summer is such a busy time for mechanics. The financial consequences were too overwhelming to even think about. We were already stretched beyond our limit to finance this voyage, because we had given up most of our charter season for this venture. We only had one three-week charter booked for the Inside Passage leg of our journey, and now it looked like that would be impossible. With the cost of rebuilding the engine from scratch, we could end up deeply in debt. So to add to my dark imaginings of being lost at sea and having to resort to cannibalism, I could now add the prospect of a future life of financial destitution and begging on the streets.

Cast Off Your Bow Lines

To top all this off, I started rehearsing in my head all the setbacks and betrayals we had experienced in our life as pastors, and I pretty much came to the conclusion that God was fed up and done with us and ready to finish us off with a burial at sea. Dark can be the contemplations of those hours between four and six lying in the bowels of a boat with an ailing motor in the middle of the ocean.

I knew I needed some perspective. Hauling my depressed carcass out of the berth, I climbed on deck, where Lyza was sitting in the helm seat on watch. Her welcoming smile dropped at the sight of my expression.

"What's the matter, dearheart?" she asked in that sympathetic way she has when somebody is in need.

I dropped on the bench and put my head in her lap. She stroked my hair for a bit and waited for me to tell her all about it. She's always been good at the emotional validation thing. I managed to convey the gist of it, including all the catastrophic imaginings (the corner of her mouth twitched into a little smile) and the more rational possibilities.

"So what is our reality?" I asked her plaintively as I ended my tale of woe. She's good with the initial empathy but also with speaking the truth in a blunt sort of fashion.

"The reality is ... this is a sailboat. It sails when the wind blows. People did this for centuries without motors and electronics."

"Yeah," I interrupted, "but we could get stuck in the doldrums, and what about our finances? If we have to head for Seattle, we'll be ruined."

She stared me down. "So what I hear you saying is that even though God has been faithful to us all our lives, we should now assume He wants to wipe us out."

"Well, when you put it that way," I conceded.

"Come on, dear. You need to get a grip on yourself. We'll just do whatever we have to do. There's no use fretting about it. You know—worry is like a rocking chair—"

"All that motion but going nowhere," I finished. She's such a little squirt to always be putting me in my place. "I think I'm ready for coffee."

A little later that morning, I turned the ignition key with my heart in my throat and was reassured to hear the loud, customary *harum* and rumble of the engine. We ran for thirty minutes, turned it off again, and checked the oil. It was still at the top mark! We tested it several times over the next hours and never lost a drop. Evidently my diagnosis was correct, and the crisis of confidence in our engine and our future was over. Captain Bluewater was back!

162

Captain Bluewater

PPSS

"The problem is not the problem.
The problem is your attitude about the problem.
Do you understand?"

Quote from Jack Sparrow

Captain Bluewater

Cast Off Your Bow Lines

The Captain's Lady /Co-Captain

18. Cetacean Visitations

August 7

First off, I need to tell you—WE MADE IT! Yep, we didn't end up floating around aimlessly after all; in fact, we made the voyage back in excellent time—only twenty-one days to landfall in Sitka on August 2 (a whole two days shorter than the way over). I can't believe we did it. We sailed right across the pea pickin' ocean, for heaven's sake, and we're feeling pretty proud of ourselves. As I write this, good, ol' *Porpoise* is steady as a rock and motoring up a calm, misty channel on her way to pick up Julie, Kai, and Chris's mom, Carol, in Juneau, Alaska, where they are flying into today. We've been having blissful sleeps in quiet anchorages now for five nights, and am I ever glad to be here.

Let me fill you in on our final ocean days as we soldiered on in the cold fog after our engine travails. For seven days after the tropical trades, we travelled through thick ocean fog with almost zero visibility, getting colder and colder as we moved further north. We were back into wearing full layered clothing with rain gear over top as though headed for the Arctic Circle. The wind kept up steadily on the starboard tack for most of the way, which was very unusual. This indicated our winds came principally off the North Pacific High, meaning we'd escaped the usual depressions that run along these latitudes.

We thought the fog would never end, and it never did completely until it turned more into a drizzle as we closed in on the coast. In the fog, we felt like we had drifted into some whitened twilight zone from which we would never return, and we began to wonder if there was any other world than our little boat capsule lit by the twelve-volt bulbs and warmed by the little diesel heater, which became a very beloved piece of equipment. With our world reduced to almost zero visibility, we became concerned about nearing shipping lanes where huge freighters plied their trade and could plow over top of us. Such a scenario obviously didn't

Cast Off Your Bow Lines

occur, since you're receiving my report, but we did almost get run over by a behemoth of a different kind.

About five hundred miles off the coast, we were approached by a fin whale. Fin whales are the second biggest whale after blue whales. We had never seen one before and had no idea what kind it was until we looked it up later in our marine mammal book. These monsters can grow up to ninety feet long and can weigh over one hundred tons. They can be easily identified by the asymmetrical colouration along their jawline: dark on one side, white on the other. All we knew was that this guy looked humongous.

He charged up from behind and made sport with us for almost an hour. He gamboled about the bow like the porpoises had. Then he sounded, flipping up his tail and disappearing from view for a couple minutes. Just when we thought he was gone, he surfaced ten feet off the boat with a great blow like a steam engine, spraying us with his watery, stinky breath. The first time this happened it startled Lyza so bad she peed her pants (not much, just a little). I asked her permission to tell you about this and she consented. C'mon, it might have happened to you too. The next time he did the exact same thing, but he also included the move of rolling sideways, just enough to fix us with his glassy eye. We didn't know whether to be thrilled or petrified. I honestly don't think this fellow meant us any harm, but I was quite concerned that his reckless behaviour might take out our rudder or prop.

I figure this whale must have been a teenager, as he was only slightly bigger than *Porpoise* (on the smallish end for a fin whale), but he was full of exuberant energy. According to the book, these massive creatures, who actually look sleek in the pictures, are one of the fastest whales, travelling at speeds of up to thirty miles an hour. He definitely was showing off his speed for us, rushing up from a half mile back and then darting under our hull to pop up on the other side of the boat, dropping back until all we saw was the spume from his breath way back. On the last of these forays, he zoomed back up to us with a friend in tow, and they both kept pace with us. Finally they charged off over the horizon to do whatever whales do, and we were left stunned, with some cool video footage to show for it.

Earlier on we had a number of other cetacean visitations. Chris and Lyza up at the bow heard the familiar chuff sound that signaled porpoises. Now porpoises weren't a rare sight on our journey, but Lyza always responded as if it were the first time she'd ever seen one. No exception this time. She immediately leaned over the bowsprit to catch sight of them, knowing they were likely going to party in our bow wave.

Cetacean Visitations

"They're Dall's porpoises!" she exclaimed to anyone who was listening. Dall's porpoises are the black and white species that look like chubby little orca whales. They're a familiar sight on our coastal cruising routes in places, but we hadn't seen any yet in the ocean crossing. A few days earlier, we'd seen a huge group of dolphins when we were still in tropical waters (that seemed a very long time ago already, bundled as we were now in layers of heavy weather gear). What a sight that had been, just like the *Planet Earth* series we loved to watch on BBC. More than a hundred dolphins came racing up behind us, and they came to play. We were enchanted by them for almost an hour. They rolled and leapt in the bow, shooting ahead and doubling back for another go in the wave cut by our prow.

Chris lay out flat on the bowsprit that day, stretched out as close to the waterline as he could get, and reached out his hand, hoping for one magic touch of a wild dolphin. They would swoop tantalizingly close but always kept that tiny cushion of air between hand and dolphin hide. When they finally broke away to find some other entertainment for themselves, we felt a little lost and lonely without them.

This time there were only a handful of them, but any porpoise is a good porpoise in Lyza's books, and she watched them avidly from the bowsprit. However, she soon pointed out to Chris that their behaviour seemed a little odd. Instead of leaping, diving, coming up on the other side, and doing it again in their usual dance, they kept veering off to the left. They did the same thing over and over for quite a while, and it was a bit of a mystery until Lyza came back to the helm to tell me about their strange movements. As she looked at the GPS screen, she saw a red triangle that had come from behind us and was about to cross our path only about half a mile ahead. She checked the screen's info box and saw that it was a freighter bound for Asia. Her mouth dropped open. "I think those porpoises were trying to warn us and steer us away from that freighter's track."

We'd all heard the stories of dolphins doing this—you know, saving the swimmers from the sharks, that sort of thing. "But," I reasoned, "we aren't in any danger from that freighter. I've seen it on the tracker for a long time."

"Yeah," she shot back, "but the porpoises don't know that!"

Cast Off Your Bow Lines

PPSS

"Can you pull in Leviathan with a fishhook or tie down its tongue with a rope?... Will it speak to you with gentle words? ... Can you make a pet of it like a bird?

—Job 41:1, 3b, 5a

Fin whale visit five hundred miles off the Alaska coast

Book Four

Explore ... Dream ... Discover

A Note to the Reader

After our ocean crossing, we made landfall in Sitka, Alaska, and went from there to Juneau to pick up Julie and Kai once again for our exploration of southeast Alaska. It was surreal to contrast this rugged icy north with the warm, palm-lined tropical isles we'd been sailing only three weeks earlier. In Glacier Bay, we watched from the safety of *Porpoise*'s foredeck as a succession of five grizzlies, one by one, lumbered out of the woods to tear huge chunks of blubber from a rotting humpback whale that had washed ashore two months before. In another spot in this huge bay, the boys took very quick dips in the freezing water, surrounded by the bergy bits of the calving glaciers.

Along our route south, we also lounged in the hot spring pools on Baranof Island, the most memorable one just beside a rushing river. When we turned into lobsters from the almost scalding pool, we cooled ourselves instantly by dipping into adjacent basins of rock carved out by the freezing torrents of the glacier-fed river.

If I were asked for a one-word impression of our Alaskan adventures, it would be LARGE. Everything seemed large—the size of the mountains, the halibut and crab we caught, the size of the giant sea lions, the distances we travelled from one stopover to another, even the size of the torrential downpour in Ketchikan that lasted five days.

As we trekked down the coastline, we reached the more familiar area of the north coast of British Columbia. The highlight there was a rare and brief Kermode bear encounter of our own on Princess Royal Island in Fraser Reach. We continued to cruise down the Inside Passage until at last we reached our own cabin in Jervis Inlet, where so many of our boating adventures began.

Once back in our natural habitat, it was hard to imagine we had actually done it. Our voyage almost felt like a dream, or something someone else might have done. But no, it was us on our little ship discovering the world we live in and exploring up close and personal a fantastic piece of the planet. Just as our surroundings had undergone an amazing transformation during that time, so our capabilities had enlarged, and our inner horizons had expanded greatly.

A Note to the Reader

In one summer—from the tropics to the glaciers

Chris and Julie in Baranof Warm Springs

19. Legacy

As a child, I took my surroundings and heritage for granted, as do most people. Growing up in our Mercer Island home on Lake Washington, my familiarity with boats and the marine environment developed from an early age. My siblings and I learned to swim like fish off the dock in our "front yard," playing water tag for hours. We learned to row dories and puttered around in them in the long summer days, fishing for bullheads among the rocks of the bulkhead and then trout, perch, and chub off the dock. When we got a bit older, we mastered the fundamentals of sailing in those same dinghies by rigging them with sails. Waterskiing became one of our favourite sports, and as adolescents, we spent hours every day charging around in our family's ski boat. On top of that, every summer my parents piled all six of their children plus the family dog on the *Nor'wester*, a fifty-two-foot motor schooner, and took off for a month of exploring, fishing, and beachcombing on the British Columbia coast. I had no idea what a rare privilege it was to have these kinds of experiences.

The cruising tradition in our immediate family started early in the twentieth century with my grandfather, Fred G. Clarke, who died before I was born. Fred had his own "cast-off-your-bow lines" moment when he left the security of his family in Iowa and moved out to the west coast. He ended up on a rural island in Lake Washington that lay in the lap of Seattle and on the doorstep of Puget Sound. Seattle was the perfect city to set up his new law practice, and Mercer Island was the perfect place to pioneer a new family homestead.

At the turn of the century, only a ferry connected the island to Seattle until the floating bridge was finally built in 1940, so boats were in the bargain right from the start. Grandfather Fred, an avid outdoorsman and hunter, began to lay plans for exploring the rich marine environment that surrounded him. He got

Legacy

the boating ball rolling by acquiring a small cruising boat called the *Gigi I* to begin his tentative explorations.

As the years went by and his law practice became successful, he had the *Nor'wester* built in 1931, which he kept moored to the dock in front of his house. His two sons—George, the older one, and Fred Jr., my father—grew to manhood in this environment and both became lawyers in their father's practice, as well as competent sailors on the family boat.

In the same tradition, my dad, who was nicknamed Ted, later had a small wooden sailboat built that he named *Fred Free*. While he was painting *Fred's* hull in the Seattle dockyards for its initial launch in 1941, news of the bombing of Pearl Harbor came over the radio. Dad went off to join the war effort, married my mother while on leave in 1942, and they had their honeymoon on board one of *Fred Free's* earliest voyages. Three sons were born in quick succession—Teddy, Tommy, and Nile—and they began packing them all into *Fred's* tight quarters for summer vacations in the San Juans and the Canadian Gulf Islands. When #4 son, John, came along in 1950, they knew they had to move up to the roomier *Nor'wester* to preserve everybody's sanity. After that move, the family boated every summer on the *Nor*, in time adding a sister, Cindy, another brother, Charlie, and a succession of yellow labs.

Fred (the boat) languished at the dock in front of our home on Mercer Island for a few years after the family shifted to the *Nor'wester*, until my older brother, Tom, and I realized in our teenage years that he was our ticket to cruising independence.

Eventually five out of six of my parents' offspring carried on the boating tradition in varying degrees on *Fred Free*, piling spouses, children, and various dogs into his very confined spaces. Lyza and I first took our honeymoon and then cruised with all our children for thirty-three years on the poor old boat until we acquired *Porpoise* in 2005. I'm almost embarrassed to admit that in my seventy years I have not missed one summer of cruising our marvellous coast.

Cast Off Your Bow Lines

Nor'wester (family cruising in the 1960s)

Fred Free (on a Dave and Lori cruise)

Porpoise and *Ariel* (buddy boating in the Gulf Islands)

Legacy

Through the years, I tended to attribute most of this cruising legacy to my father's side of the family, but in the last couple of decades, I've learned that much of this heritage also came from my mother's side from centuries back. Mom's maiden name was Vinal. This somewhat eccentric moniker was placed in the middle of my name, John Vinal Clarke, as a tribute to those family roots. I tried to keep that odd name a secret in school because my classmates, once they unearthed it, amused themselves at my expense with comparisons to floor coverings (vinyl) and men's washroom facilities (urinal).

However, later in life, while on a search for my roots in New England, I discovered that my unusual middle name had an illustrious history in early America, dating back to a certain Anna Vinal. She cast off her bow lines, leaving England with her husband, Stephen, on one of those leaky *Mayflower*-type ships. Landing in Massachusetts in 1630, this couple begat three little Vinal boys before Anna was widowed. She raised those boys to become the progenitors of three prominent Vinal lines that can be traced through pages in the annals of American history.

Stephen's ancestry could be traced all the way back to 1430, where it appeared as the ancient family of Vine Hall. John Vinall of that line (good name) was granted a coat of arms under Oliver Cromwell in 1657. Suddenly bearing my previously embarrassing middle name had a whole new shine to it. There's even a good-sized island off the coast of Maine named Vinalhaven in honour of one my ancestors. Of course, it was too late to get satisfaction with my high school friends by doing a little name-dropping, but these discoveries made me curious to find out more about my lineage and to chase down some family history.

I was already aware of two whaling captains in my mother's family tree dating back to the 1800s. I'd heard Skipper, my maternal grandfather, tell some stories about them. He told us that one of these captains, William L. Vinal, sailed a whaling ship out of New Bedford in the 1850s, and his painted portrait still hangs in the New Bedford Whaling Museum. The other whaling captain, Charles Veeder, hailed out of Nantucket. He had made a fairly late-in-life whaling voyage to secure his daughter's financial future. Tragically, he had been shipwrecked, lost his whole cargo, and died, broken-hearted, in Tahiti. I never had any cause to doubt this story until many years later.

This version of the story remained part of our family lore until a Nantucket writer contacted my sister concerning a book she was writing about Charles Veeder's wife, my great-great-grandmother. Apparently, Susan Veeder had the rare distinction of having travelled for four years with her husband on his whaling ship.

175

Cast Off Your Bow Lines

On board she had kept an account of life on a whaler, giving a woman's point of view in a mostly man's world. It was unusual enough that she had accompanied her husband, because wives rarely did, but it was even more unique that she kept a journal of her journey, complete with her illustrations. Betsy Tyler, the author from Nantucket, was in the process of writing a book called *A Thousand Leagues of Blue*, which has since been published and is available through the Nantucket Historical Association. The book gives a fascinating firsthand look at the arduous life of a whaling family in those times.

Because of the research she had done for her book, Betsy was able to both fill in some details and correct some impressions we had about our ancestors, Charles and Susan Veeder. The Veeders lost two of their three sons at sea when they were washed overboard as young men. They also had a daughter, Marianna, born years after the boys.

Supposedly it was for this daughter that Charles made one last whaling voyage. This is where the story diverges from what had been passed down to us. According to well-documented sources, Betsy discovered that the shipwreck was a myth. Once the captain reached Tahiti, he left off chasing whales and started chasing Polynesian women instead. He became enamoured of a particular woman and the ways of the South Pacific. Abandoning the responsibility of the ship to his underlings, he became a roaring drunk, spending most of his time ashore. When he did come back to the ship, he brought women and drink with him and made an utter fool of himself with his debaucheries. After also abusing his crew terribly, he was finally reported by his men to the local authorities and relieved of his command by the ship's owners.

The circumstances of the rest of his life in that part of the world remain shrouded in mystery, except that he died there in 1878, possibly swept away in a hurricane. It's uncertain how much of this sorry tale Susan Veeder became aware of in her time, but it's clear that the real details weren't passed down to my grandfather's time period. Learning of his ignominious end dispelled any idyllic illusions I might have had about my illustrious ancestors. I suppose there are a few skeletons in every lineage closet.

Despite his disgraceful downfall, my research into the family tree revealed a proud tradition of salty ancestors and confirmed that the sea genes in my blood came from a long way back. Charles and Susan Veeder's daughter, Marianna, grew up and married Charles, the son of William Vinal, the New Bedford whaling captain, linking up the two whaling families at this stage of history. Those

176

Legacy

seafaring genes passed from both my great grandparents, Charles and Marianna, to my grandfather, Elwyn (Skipper) Vinal.

Captain William L. Vinal in New Bedford, circa 1850

Charles and Susan Veeder in Nantucket, circa 1850

Skipper eventually married a New Bedford girl, Grace Tripp, and whisked her off to the west coast, eventually settling down the road from the Clarkes, making my grandparents on both the Clarke and Vinal sides pioneers on Mercer Island. There he and Grace raised their son and three daughters. One of his daughters was named after Marianna Veeder but was called Nuky all her life. She raised her family on Mercer Island too. Two of his daughters, Helen and Cynthia

Cast Off Your Bow Lines

Lee (my mother), married the two Clarke boys, George and Fred. They had nine children between them, making us all double cousins. All of these families lived in close proximity, with *Nor'wester* and *Fred Free* moored respectively at the docks on their waterfront lots. For many decades, both boats were iconic features on the island and recognized by boaters and islanders alike.

Thinking back on the life of Susan Veeder, I can't help but make some connections in my mind between her and my mother. It was uncommon in those times for a woman to travel with her husband on a whaling ship. It was even rarer that she kept a journal. To hear a woman's voice from the man's world of the whaling industry in the late 1800s was very special indeed. To discover that a book had just been written primarily around her journal sent a thrill of discovery through me. No wonder my mother was the one to initiate the unique log-keeping tradition in our cruising clan. No wonder she was such a game sailing partner with my father and built such a rich culture of adventure and discovery of the natural world into our summer sojourns. I think Susan's pluck and creativity percolated down into the genetic pool to seep out two generations later into mother's embryonic juices. I'm certain they leaked into my siblings' and my genes as well. For this reason, I dedicated space for mother's cruising log of 1952 in the final chapter. Her log journal reminded me of what a great part of the warmth and colour of my childhood memories upon the water are due to her.

Those sneaky genes also flowed down the feminine side into my little sister. Cindy carried on the plucky mariner heritage, becoming a teenage skipper of *Fred Free* and spending summers with a crew of her beautiful friends. She married into the Peterson clan, another Mercer Island waterfront family, that cruised in the summers on their own large sailboat. Dean, her husband, acquired and renovated a huge old tug called the *Gilspray*, on which they and their family of four children cruised for many summers.

At times all the boats would meet up somewhere in the coastal islands or on the dock in front of our Mercer Island home as a great family flotilla. When parts for the old tug's engine became obsolete, *Gilspray* transformed into a bed and breakfast on Lake Union in Seattle. Cindy and Dean are still cruising on a Kadey-Krogen appropriately named the *Wester Lee* (combining the name *Nor'wester* and the name *Lee*, as my mom was called). Carrying on this hearty female tradition into the next generation is my daughter, Julie, who sails every summer with her "boys," as she calls them—her husband, two sons, and Rumi, their male Portuguese Water Dog. They cruise on *Ariel*, their spritely C&C sloop. When we

178

Legacy

meet up to buddy boat, she looks like *Porpoise*'s offspring in her matching green and white colours.

It's obvious that I owe much to my forbears for the unfolding story of our Pacific Encounters, and I aim to keep passing on these priceless traditions to my own progeny and others who get bit with the same bug. Now in my seventieth year, with the final port of call somewhere on the horizon (although thankfully I don't see any land yet), it's important to have future captains in the pipeline to secure this sea-loving heritage. Our son, Matt, and son-in-law, Chris, who crossed the mighty ocean with us, clearly have earned title to seaworthiness. They're already certified coastal skippers under the auspices of the Canadian Recreational Yachting Association (CRYA) and plan to become sailing instructors. We also have a number of budding sailors in the next generation down, our seven grandchildren. When asked about their favourite place on earth, most of them can't decide between the cabin in Jervis Inlet or *Porpoise*—except for Jude, Julie's younger son, who always refers to their boat as "my *Ariel*," and at only eight years old will sit up at the bowsprit for hours just watching the sea and sky.

I'm not sure when the next generation will be able to convince the old seadog to pass on to them the helm of *Porpoise*. Not until he's good and ready, for sure. By the time he's eighty, I suspect they'll probably conspire to pry their dad's hands off the wheel before he sinks the ship or somebody else's fancy yacht. I'm not really worried about it, because I know they're able men and women seafarers. As my father used to say, "The generations shall not fail."

Cast Off Your Bow Lines

PPSS

    Lives of great men all remind us
    We can make our lives sublime,
    And, departing, leave behind us
    Footprints on the sands of time;
    Footprints, that perhaps another,
    Sailing o'er life's solemn main,
    A forlorn and shipwrecked brother,
    Seeing, shall take heart again…

                                Excerpt from "A Psalm of Life"
                                By Henry Wadsworth Longfellow

My mother, Cynthia Lee (Vinal) Clarke, in 1944

Legacy

My Mother's Creed

The photo on the previous page includes my mother's creed that she wrote when she was only sixteen.

1. Try to live up to this creed.
2. Be natural in every way, not artificial.
3. Be thankful for what you have.
4. Lead an active, interesting life.
5. When you make a promise, keep it no matter what.
6. Always be a good sport.
7. Keep yourself healthy.
8. Don't criticize other people. If you don't like them, keep it to yourself.
9. Help other people whenever you can.
10. It doesn't hurt to be polite and courteous.
11. Don't be disagreeable and argue. It will never get you anywhere.
12. Always be loyal to the members of your family.
13. Don't give false impressions.
14. Give as well as take.
15. Be understanding.
16. Be tolerant and patient.
17. Try to understand the psychology of young children and help them along.
18. Try to leave things better than you found them.
19. Be able to take criticism.
20. Be generous with your belongings.
21. Save your money for a rainy day.
22. Be able to lose cheerfully.
23. Be able to sacrifice things to help others.
24. Try to do things a little better than average.
25. Be a good student but not a drudge.
26. Be able to laugh and take a joke.
27. Be able to be emotional at times and admit you are human.
28. Stick to the truth always.
29. Be sincere in your friendships.
30. Be frank.
31. Live each day as it comes along.
32. See beauty in everything around you.

Cast Off Your Bow Lines

My sailor sister Cindy in captain's hat, 1973

Legacy

My daughter Julie and grandson Jude on *Ariel,* 2020

20. Princess Encounters

Thirty-five miles up Jervis Inlet from our cabin is Princess Louisa, a place many would call the crown jewel of British Columbia's coastal inlets. Our extended family developed a special connection with the Princess over six generations and more than a hundred years of cruising the west coast. This is the setting where my grandfather began his cruising life in 1927. The same place where my father and mother boated in the forties and fifties, first as a couple and then with their brood of five boys and a girl. The same place where my two-year-old self sat in a clam bucket on Mac's float, and fifteen years later met the sweet seventeen-year-old who became my wife. The same place where we cruised time and again with our own family of four and later with our grandchildren throughout their entire childhoods. Finally, it was also the place where we took scores of people from all over the globe for their own Princess encounters as charter guests on *Porpoise*.

Charter Cruise in August, 2015

I swung the bow around the native burial island and toward the log structures and totem poles of Malibu. Then I pointed the bowsprit directly at the centre of the fort-foot-wide channel. Usually my timing was good to hit these rapids at slack, but today we were late, and the current had picked up speed and was running hard. As a responsible skipper, I've learned when to yield to these forces of nature, but I have many years of experience with this particular set of rapids, so I made a quick assessment of the conditions. Doing so gave me the confidence that, although this passage might give my guests a thrill, it would still be within the bounds of safety for ship and crew.

The boat shot through the tightest part of the narrows to a frothy vortex on the far side. The inflow of a current station is often the smoother, calmer section, and the outflow side is the more turbulent one that can make for a crazy ride. For a few seconds, *Porpoise* churned and twisted in the powerful whirlpools before

Princess Encounters

the swirling countercurrents suddenly released us from their grip into the flat, settled waters of one of our favourite destinations.

In the charge through the channel, our crew hardly had time to appreciate the Malibu Young Life Camp, filled to the brim with more than three hundred exuberant teenagers cavorting in the pool, playing in the sand volleyball court, and lounging all over the moss-covered bluffs. Our crew planned to come back and visit the camp the next day when the rapids were flowing out. Then the timing would be right to take a tour of the place and sample the homemade ice cream.

At the moment we were more interested in ushering our guests up the four miles of serene, emerald inlet to the head, where thundering Chatterbox Falls fell 120 feet to the sea. The surrounding evergreen-clad cliffs and massive granite rock faces climbed abruptly to seven-thousand-foot glaciers that enclosed this seemingly secret Shangri-la. These sights cast a mystic spell of wonder over anyone we brought here. It was especially gratifying to bring the ones who came from the flat Midwest states or the Prairie Provinces. The fjordland country of Jervis Inlet was such a fantastic contrast to their usual views that we never had to worry about them experiencing boredom or having a ho hum remembrance of their holiday.

The folks we had aboard this particular week were from Calgary—a father, mother, two adolescent daughters, and an eight-year-old adopted son from Ethiopia. What a treat to usher them into our secret marine paradise of blue waters warm enough for swimming in late summer, oysters and clams on our cabin beach, and long broad reaches sailing up the mountainous length of Jervis Inlet. I knew this trip would become a treasured memory of their family's life, a segment of their journey where they drank deeply of an elixir that only our Pacific Northwest coast can provide. It would be a chapter in which their boy sat on the bowsprit, his arms spread wide and a huge grin plastered on his face. It would be a chapter in which their thirteen-year-old girl latched on to a fifteen-pound lingcod off Patrick Point and screamed her brains out in fright as its huge toothy mouth emerged out of the dark depths to greet her. It would be the time her younger sister showed her up by stepping up to land the fish and hold it up by its gills for the photo op.

Their mom and dad, after fifteen years of marriage, would pull out their original wedding dress and tuxedo from their luggage and renew their vows in front of our cabin site as their grinning son hopped around excitedly on the rocks beside them. They would keep forever the memory of their daughters, in matching purple satin dresses, singing in celebration just as the sun set in a

185

Cast Off Your Bow Lines

pink glow behind them. All of this would be their unique and precious pacific encounter. Their trip marked our seventh cruise that summer passing through Malibu rapids, quite a few times even for us in one season, and this was just one year of voyages.

The previous passages of our family reached all the way back through the generations and years to the 1920s and 1930s when my grandfather and great-grandfather cruised …

Charles Clarke's Cruise in August, 1931

The high pointed prow of the *Nor'wester* plowed the still green surface, not unlike the plows that cut the dark soil of the Iowa farm Charles had lived on from his earliest recollections. Everything else that met his eyes in this fantastic landscape was so radically different from Iowa that he had trouble holding on to its reality. It almost seemed as if the stupendous mountains surrounding him were paintings on the huge backdrop of a stage, were it not for the pervasive and exhilarating fragrances of saltwater shores and mild northwest airs filtering down through the damp evergreen slopes. Charles was completely enchanted. He never moved from his perch on the stern bench for the entire forty-mile journey up into the mountainous reaches of Jervis Inlet, right to the rapids at the entrance to Princess Louisa.

On board with him for the *Nor's* maiden voyage was an interesting assemblage of personages, all come to visit this coastal wilderness in the latter part of the roaring twenties, just before the depression years would prevent most people from even dreaming of doing yachting excursions. First was the weighty and all-consuming presence of "the Governor," Charles' father, who came all the way to the coast to be with his oldest son, Fred, and see this vessel his firstborn had built with the earnings of his newly-established law practice in Seattle. Everyone was proud of the accomplishments of Fred and his wife, Mabel, and admired their two strong sons, George and Fred Jr, also on board for this trip. Even Nile Kinnick, their cousin, was crammed into a berth in the bunkhouse-style aft cabin. This was to be a memorable family event and the beginning of generational pilgrimages to Princess Louisa that had already begun and would continue for a hundred years.

Charles, the younger son, tended to stay in the shadow of his illustrious and larger-than-life father, who was governor of Iowa from 1913 to 1917. This was partly because of a retiring and contemplative nature, but also due to the fact

Princess Encounters

that from a very young age, he'd suffered from severe arthritis, which left him practically crippled and prone to more sedentary and literary pursuits. While his cousins talked excitedly among themselves of the fish they would catch and the cliff jumps they would make, Charles pondered the mystic forest and the possibilities of building a more permanent retreat on the shores of these quiet reaches.

Upon his return to Iowa after this cruise, he was so moved by his experiences of the great coastal wilderness that he sat down and wrote a poem to capture his many poignant perceptions and the sentiments they evoked in him. Little did he know then that in so doing, he would become a sort of family prophet. This poem remained hidden in a file somewhere in my parents' attic until one day in the mid-nineties my dad pulled it out to show to me. At that time, the two cabins built by my brother's family and my family were under construction. In 1999, copies of his poem were hung in places of honour on the walls of both cabins built along the shores of Jervis Inlet by the great grandsons of the Governor. (This poem is also in a place of honour at the end of this book.)

Charles never made it back to this coast of his "solemn reverie," but it obviously made a deep impact on him. His brief encounter stayed with him and his tribe through the following years of depression and World War II. His nephew on that cruise with him, Nile Kinnick, after winning the Heisman trophy for college football in 1939, would die at twenty-two after ditching his plane within sight of his aircraft carrier in the Pacific theatre. Nile's brother, Ben, was shot down with his plane and perished in the Coral Sea. Others of the Clarke clan would live long fruitful lives, some in the Midwest and others within sight and sound of the Pacific waterways. Whether during short or long visits to the coast, the Clarkes were never quite the same after exposure to these mountainous inlets. The pioneering inroads made by my grandfather Fred in the twenties and thirties would be traversed over and over again by subsequent generations on the *Nor'wester* and on other family boats, including *Fred Free*, the *Gilspray, Ariel,* and our beloved *Porpoise.*

Although there are undoubtedly many unrecorded family pilgrimages to these parts that I know nothing about, a few visits were captured, not just in photos, but in the much harder-to-handle medium of print. Although it's said a picture is worth a thousand words, sometimes a thousand words can say much more than a picture could ever convey. One such account was creatively written by my mother in 1952 and is included here because it captures so well her humour and spirit, as well her connection to cruising, to nature, and to Princess Louisa.

Cast Off Your Bow Lines

My father Ted and four sons in 1952

Lee Clarke's Cruising Log in July, 1952 (an excerpt)

July 1

We're starting this year's cruise on the first day of July and expect to spend nearly a month with just our own four boys cruising the waters of the San Juan islands, the British Gulf Islands, and as far north as Jervis Inlet and Princess Louisa. Our new house, only partially built, and a yard full of piles of dirt and lumber are excellent excuses for a vacation, if excuses are necessary.

After a week of harried preparation, we were finally ready to leave at 6:00 a.m. Tuesday morning, July 1. Teddy, nine, is quite a help as deckhand this year and can handle the bow line in the locks, can coil ropes, and can run the outboard, as just a few of his capabilities.

Princess Encounters

Tom, seven, and Nile, five-and-a-half, are easy to watch and content to take each pleasant day as it comes, extracting the most possible enjoyment from it. All three find each anchorage a delightful new challenge and a new source of crabs, starfish, pretty rocks, driftwood, and wildflowers to be collected and saved until we return home a month hence.

John, twenty-two-months-old, isn't quite so easy to control. Everything centres around J.V. and his activities. We may be cooking alive inside the boat, but we can't open the doors because John will fall overboard. We pull up at the locks and our first duty is to hook J.V.'s harness onto the cabin door, put on J.V.'s life preserver, extract J.V.'s fingers from between the boat and the wall of the locks, and comfort J.V. when he howls because he thinks we're all going to get off at the locks and go home, leaving him in charge of the *Nor'wester.*

While the boat was tied up at Bryant's Marina to take on gas and water, John ran along the catwalk, sat down on the edge, and fell over backwards. Ted grabbed a wrist and I grabbed an ankle just in time to save him a cool dunk in the lake.

July 2

Everyone was up bright and early exploring the beach and demanding pancakes for breakfast. John was wading in the nude by 10:00 a.m., and I had a wonderful chance to clean the boat unmolested. Later, Tom, Nile, Mishka [our Samoyed], and I climbed a long hill by the beach in search of wildflowers. We found several lovely species, which we identified on our return by looking them up in a little wildflower book. We found the hillside carpeted with orange tiger lilies, Indian paintbrush, fragrant yarrow, daisies, wild onion, and a darling purple bell-shaped flower we were unable to identify.

We played on the beach until four o'clock and then took off for our Puffin Island anchorage at Matia. By eating dinner on the way, we were able to have a seagull expedition before bedtime. We thoroughly explored Puffin Is. and found the baby seagulls hatching out all over the island. Nests were everywhere, and we had to be very careful not to step on babies hiding in the tall grass covering the island. It was like an Easter egg hunt to the children, with John in the lead saying, "See, boodie!" The babies were difficult to pick up because they invariably spit up the herring their parents had just shoved down. All the shags [our family's name for cormorants] were still sitting on eggs, however, so we didn't even bother them.

Cast Off Your Bow Lines

July 3

Ted was up at 6:30 fishing, but the rest of us slept in until 7:30. Nile, as usual, had fallen out of his bunk again but failed to even wake up, so when he awoke he found himself on the floor and was quite surprised. We left for Sucia immediately to dig clams, and Tom helped ... or rather started to help, until he discovered some fascinating and odiferous specimens to add to his collections.

By 11:00 a.m. we were heading for Sydney. On the way, all four boys took a good nap. We spent a wonderful hot afternoon on Sydney spit with John cavorting naked all over the sandy beach and having the time of his life. We ate wieners, spaghetti, and salad on the beach but found everything coated with a layer of sand, which J.V. had tossed in the basket at some time or another. John kept trying to roast a wiener on a stick but it kept getting stuck in the sand. The final blow was when he sat down in the salad bowl and tipped the mustard jar over in the sand. Ted and I finally brought him out to the boat and put him to bed while the other boys played cowboy on the beach. When little troublesome was finally asleep, we sat by a large beach fire under a full moon, and Teddy and I played the accordion.

July 4

No trouble getting up early this morning! The boys with their fireworks were ashore celebrating the fourth long before breakfast. At noon we had a large clam bake on the beach, and then all the kids took very long naps so they could keep awake until dark. We took a long walk around the point, south along the spit, and John had a great time. However, "walkie" to him means ride on Mama's sunburned shoulders. We finally had trekked so far we sent Tom back for Ted and the motorboat to come and get us. Teddy and John went back in the boat while Nile, Tom, and I walked back, bringing half the beach with us (fishing floats, driftwood, peels of white birch bark, and pretty shells).

By afternoon, there was a very strong wind blowing, and it became quite rough. We decided it was a good night to sleep ashore, so we picked out a good clear spot above the tide and spread out our sleeping bags. Ted had to sleep on the boat anyway, so we were happy to turn John over to him for the night. It was a cold, blustery night, but we set off our night fireworks and then crawled into our sleeping bags fully clothed. The sky was quite spectacular with an almost full moon and dark scudding clouds racing across it. Our only disturbance was a persistent night hawk who insisted on sitting upon a log about fifteen feet from

190

Princess Encounters

us and calling to its mate across the marsh. He kept this up all night, I guess, because every time I woke up I could still hear him.

July 5

We awoke early with a beautiful blue sky and bright sunshine everywhere. It was a perfect day for crab fishing, as there was not a ripple on the water and a very low tide. We all went out in the big dory to try to catch some. Every time we'd spot one, all would rush to one side of the boat. To keep from swamping, I'd have to rush to the other side, thus losing sight completely of the location of the crab. John, becoming bored, would be trying to start the engine, and the rest of the children would be arguing over "who gets to net this one." By this time, we would have drifted ten feet away, and the elusive crab would have long since taken refuge in the surrounding seaweed. For some unknown reason, we were lucky enough to catch six good crabs, one a very good sized one and the rest medium.

Two passive Indians in another canoe were quite amused at our shrieks of "Quick, get him, over here."

"Where?"

"Down there!"

"I don't see him. Ohhh—he's huge!"

Great rush to one side while slimy bilge water sloshes over our bare feet. Then "Get off my toe!"

"Get your feet out of the net."

"Grab John; he's going over!"

"Let me net him."

"No, it's my turn!"

By the end of this sojourn, John was exhausted (he wasn't the only one) and went right to bed, while the rest of us ate crabs and clams on the beach and took naps in the sand. It was a very hot afternoon and everyone was very sunburned. We took a trip in the motorboat out to the white marker on the end of the other spit. It was a gorgeous night, and after the children were in bed, we sat on the deck and watched the full moon rise and played the accordion and sang songs.

July 7

This morning it's still very windy, but we all clambered into the motorboat and headed for our favourite clam beach. This certain point surpasses all others in quantity and quality of butter clams. The boys became very enthusiastic when they discovered how easy it was to turn them over with shovel, pick, clam shell, or just one's hands. John spent the whole time gleefully running through the holes full of water where we had been digging. He ended up flat on his tummy in the mud, splashing and wallowing around like a little pig.

Ted found the clutch had a broken bearing and was making a very strange noise, especially when he increased the speed. Rather than damage the clutch permanently, he decided not to use the big engine at all until we could get a mechanic to look it over.

July 8

When we awoke this morning, Ted was already up and preparing to leave for Nanaimo, the only large port where we could get our engine repaired. We travelled the whole distance (twenty-seven knots) with the trolling motor, part of the time bucking a strong tide and not making more than three to three-and-a-half knots an hour.

It took us from 8:00 a.m. to 6:30 p.m. to reach Nanaimo, and by that time the boys were fit to be tied. For a while they were content to play checkers, old maid, parcheesi, and dominos, while John systematically tore the boat apart, but finally they decided John was having more fun, so they joined him in creating complete shambles throughout the boat, They played "lion-in-the-cage" by taking turns hiding in closets in the after stateroom and feeding each other tidbits they had filched from the galley. The final straw was when Nile lunged head-first down the hatch and lit on his nose on one step and bounced to the floor on his back with a tremendous howl. He then proceeded to bleed all over the clothes and sleeping bags lying on the floor and dripped blood through the bathroom all the way to the pilothouse. He still has a large bulbous red nose and swollen lip but otherwise seems to have regained his composure and good humour.

July 9

The engine is fixed but the necessary part for the clutch has to be sent for to Berkeley, California, so they removed all the parts of the clutch and we are functioning without one. It seems this is very simple, as you can start either in forward or reverse, but in order to change gears, you must first shut off the engine. The main difficulty is landing at a dock or manoeuvring in close quarters, but

Princess Encounters

if wind tide, God, and Lady Luck are all on your side, your chances of getting the engine turned on again in reverse before you ram the dock are pretty good.

July 10

We left Nanaimo at 4:30 a.m. just as the sun was rising and crossed the straits by 8:45 a.m. The children slept nearly three hours of the crossing, and Tom was the only one who became seasick when we hit rough water. The weather is still absolutely perfect, and we had a wonderful run up the channel through the huge, towering cliffs and heavily wooded mountains on either side to the entrance of Princess Louisa Inlet. The tide was ebbing very strongly through the entrance, so we decided to eat a picnic lunch beside a beautiful waterfall directly across the bay from the entrance. We found a perfect spot in the cool woods by the river, whose banks were literally carpeted with several varieties of ferns, lichens, and small wildflowers. It was a very swift and deep rushing river, so we were glad it was John's nap time and that he was sound asleep on the boat. We found a tiny trail through the woods and followed it through almost virgin timber with the compost of centuries springy underfoot.

The whole forest was fused with a grey-green cloud of low-hanging moss, and the air so still you could hear distinctly the buzz of a bee or a cracking twig, which the children were convinced was the footsteps of a bear. (I thought perhaps it was myself, so we returned rather quickly to the friendly sound of the rushing river.) We found a very small ground cover that very closely resembles a dogwood in leaf and flower except it's a dicot, instead of a monocot with a whorl of four not three petals. We also found lovely, fragrant, delicate pink wild roses growing everywhere. Our only casualty was when Ted fell in the river with the camera on his arm.

By 3:30 we entered the inlet with a flooding tide and anchored within one hundred feet of the *Blue Peter* and several other cruisers. There was a beautiful large black sailboat anchored near, and we discovered that Harry and Sally Jones owned it, so we paddled over for a highball before dinner and exchanged our cruising experiences. After we got the little boys to bed, we each collapsed on a seat in the pilothouse and were snoring contentedly when we were hailed by MacDonald in his red canoe, We carried on a rather hazy conversation for about half an hour and then left our bunks just long enough to wave to him as he went in search of more stimulating company.

Cast Off Your Bow Lines

July 11

This morning we moved round the point to anchor in a quiet little cove surrounded by three of our own private waterfalls. We had to anchor very close to
shore, as it's quite deep here. This time of year there is loads of water coming off
the mountains, and the big waterfall is a really thundering cataract.

While Ted painted parts of the boat and watched J.V., the three boys and I
took a great hike through the woods following a small rushing stream. We found
lots of lush huckleberries and really ate our fill of both blue and red berries. Due
to various injuries and difficulties encountered along the way, we didn't climb too
far but came back down and had a wonderfully cool and refreshing swim in both
salt and fresh water. I finally caught up with the diaper washing and hung at least
three dozen in the rigging of the *Nor'wester.*

In the afternoon, we called on Mac on his barge and tried to catch some
small perch under the wharf. We took along John's private swimming hole in the
form of the large clam bucket, and he was perfectly happy sitting in about ten
inches of water and splashing about for all he was worth. It was so hot tonight
we had a cold supper and went to bed early.

July 12

This morning we made a great trek to the large waterfall and struggled through
roaring rapids (carrying Nile and John) to where the mist was so heavy it was like
a heavy rain. We shivered and tried to look extremely happy while Ted snapped a
picture of us, but John insisted on burying his head in my neck to keep the mist
out of his face.

Our next stop was Trapper's Rock and more huckleberry feasting. Ted cut a
long, straight pole from a green maple and fitted a spearhead to one end. With
this we managed to spear a small perch under Mac's barge, which we'll use as bait
to catch a red snapper under the cliffs.

Meanwhile, poor old Mishka, our Siberian husky, had attempted to follow
Ted in the motorboat through dense underbrush and had managed to maroon
herself on top of a sheer fifty foot cliff. Harry [a family friend] and I climbed up
on both sides of the cliff to coax her down, but she was really stuck, as apparently
a large piece of moss she had stepped on had given away and left a sheer cliff
behind as well as ahead of her. Darkness interrupted our efforts and we had to
leave her on the cliff all night. We felt very guilty as we left her wailing pitifully
for us to come back and rescue her.

July 13

Tom had a very sleepless night with a sore toe and nightmares most of the time. He's sleeping in this morning, so consequently poor Mishka is still warming a spot with her posterior on a narrow ledge on the cliff. It will require the whole family giving advice to rescue her, I'm sure.

Ted took the other boys to Echo Cliff to use our perch bait and couldn't even get the line down before a good-sized rock cod was hooked. They rushed back immediately to spear some more bait, so it looks like we'll be having a nice fish dinner tonight—a welcome relief from Franco-American spaghetti, since we have now run out of everything both palatable and fattening.

Getting Mishka off the cliff was just another addition to our harrowing but ludicrous experiences of this trip. With Mishka looking drawn and harried from her lonely night on the cliff, we approached her from both sides. From a perch above and to the left of her, I was able to work down the cliff to her, tie a large rope around her even larger avoirdupois, and throw the other end of the rope to Ted, balanced on a cliff to the right. While Ted pulled, I pushed, and with a great undignified lunge, she almost reached safety when the rope slipped off and she slid off down the face of the cliff, landing with a great crash in some bushes along the edge of the water. The only problem then was to get Mom off the cliff, which was soon accomplished with less ceremony and more dignity(?) by scrambling back up the cliff the way I had descended.

About 5:00 p.m. we moved the *Nor'wester* to Echo Cliff and put the small boats in to fish. John stayed on the big boat with me and was thoroughly fascinated in going "hoo-hoo" and hearing his echo come back. Of course half the time it was not his echo but his big brothers on shore. Despite alluring bait, they were not biting, and the boys came back empty handed. Oh dear—spaghetti again!

July 14

We left bright and early this morning to go through the pass at slack time, 7:30. We had a fast run down from the inlet, except we've had such extremely high tides recently that the water is full of driftwood. Some very large logs and stumps but mostly scraps and debris. On our way out, I was steering, and when I saw a small stick about two inches across and four inches out of water pointing straight up, I thought very little about it and hardly even bothered to miss it, but as we went by quite close, I could see the large shape of a huge tree trunk attached to it underwater, perhaps two feet in diameter and six to eight feet long.

Cast Off Your Bow Lines

The straits started out to be very calm and quiet, but as we progressed, they roughened. By the time we were halfway across, we were really rolling, and Tom and Nile began to look very green, while Teddy and John went promptly to sleep and slept the whole way. Mishka and I deliberately stayed outside and kept our noses pointed into the wind while Ted steered the boat, looked disgustingly chipper, and periodically counted the inert bodies littering the decks.

We made good time to Nanaimo and spent the next hour replenishing our food supply and buying the boys milkshakes. It was terrifically hot in town, and the boys were doing considerable moaning because they'd had to put on shoes for the first time in a long while, and all had one or more sore toes to complain about.

By four o'clock, we were again heading south through Dodd Narrows. Ted discovered one of the gremlins (identity unknown) had left the bathroom faucet on, and the bathtub was full of water. I threatened to make the guilty one take a bath in it but never went through with it, since we strongly suspected Johnny of the deed.

***

The rest of my mother's log chronicles the further family adventures characteristic of many a family cruise though those years. Those month-long voyages every summer on *Nor'wester* were a centrepiece of my idyllic boyhood. My mother and father made their family a priority, and they made a priority of exposing us to this very special coastal environment. I will always be grateful that they did.

Lyza and I have endeavoured to provide our children and grandchildren with this same invaluable legacy. As is obvious by this little sample of my mother's writing, she was the nurturing centre of that wild upbringing with six children on a boat in British Columbia and at our waterfront home on Mercer Island. She and my father are gone now, having lived into their mid-eighties. *Nor'wester* lives on, having passed into other hands after some ninety years of family ownership, but the heart of my parents' rich heritage lives on with us now on *Porpoise*. As a final goodbye to her and those formative years, I include one last entry from that summer's colourful cruising log.

July 17

Ted and I decided on no naps today but early to bed for all children. Our no-nap routine worked fine, except we forgot to take into account young John. He was full of old nick tonight and kept talking (babbling) a steady stream to Teddy and Nile in the opposite bunks. Then he found it great fun to throw his teddy bear and Lena (his stuffed lion cub) across the room and yell "Uh, oh, doggie!" which in English means "Mother, get me my dolls, quickly." Finally I went in and tacked a blanket up in front of him. I heard him say, "Bye, Mama." Half a second later, a curious head pokes around the end of the blanket and a small voice says "Oh, hi, Mama."

* * *

I'm so glad my mother took the time to write this little piece. None of us should think our story too unimportant or too insignificant to record, at least for our own progeny if for no one else. Not able to see for myself my own two-year-old antics, I have here a little window of warmth—a preserved view of myself, my siblings, my mother's heart, and those summer days. She's gone now, but I remember. She has helped me do so. Thanks, Mama.

JV taking a bath on Mac's float in 1952

Cast Off Your Bow Lines

Ted and Lee Clarke (my parents) on *Fred Free* in 1972

John and Lyza in June, 1968

It was the same inlet, same place. How many times had the family come to Princess Louisa on the *Nor'wester* and yet, somehow, as a child this Malibu Camp had never registered with me. Like a lot of kids, I must have been reading comic books or something when the really good scenery and places were slipping by. I remember water skiing here in the perpetually calm water beyond the rapids. I used to ski and ski and ski behind our old sixty-horse motorboat, my dad at the wheel as I skimmed over the glassy inlet and ducked under the tree branches that hung over the water by the shore. Some of these Princess images were indelibly imprinted in my head from the multitude of framed black and white eight-by-ten photos covering the walls of our wood-panelled den in the family's Mercer Island home. There was the picture of the four Clarke boys standing on the bow of the *Nor* with Johnny in Mac's arms, Chatterbox Falls in the background. And the photo of me on Mac's float, bathing in a clam bucket. Both images were taken on the same cruise my mom so colourfully described. They were imprinted like black and white film negatives in the back of my mind and etched into my psyche as part of our collective summer tribal ritual.

Now as we chugged through the rapids on the camp's boat, the *Malibu Princess*, and tied up at the dock, the real-life images of this magnificent inlet were amplified by the glow of my own teenage perceptions. Princess Louisa of 1968 was a whole new experience for me. The camp, owned by the organization Young Life, used this backdrop of spectacular natural beauty to serve up their mixture of teen-geared fun, the cultivation of meaningful relationships, and a serious underlying challenge to consider the deeper meaning behind the beauty of the created world surrounding them.

Princess Encounters

A couple of years earlier, I'd been a camper at Malibu and had an encounter with the God who inspires the mission of Young Life. Although there hadn't been too much of an outward change in my focus or behaviours as a result, that encounter was the reason I'd been accepted on work crew, a month-long stint at the camp working for free in whatever capacity I was assigned.

At that stage of my life, the glacier-clad peaks and granite cliffs took a back seat to finding a way to fit in socially, discovering what the scene was all about here, and, hopefully, having a chance to show off my water-skiing abilities. Also, there was the hope of some Princess encounters of another kind to be had with some of these west coast girls. In that regard, the scenery was indeed something else.

Who knows what forces of nature, personality, attraction, and even destiny set to work to cause two people previously unknown to one another to connect, bond, and "fall in love"? Although summer infatuations of this sort rarely endure the deeper tests of time, when a chance meeting is made and does last, it can be a marvel to behold. Lifelong bonds are established and life-altering decisions are made affecting the course and direction of many destinies. It can even result in the creation of eternal beings who would not otherwise have existed.

To really ponder such considerations in terms of the ramifications of chance encounters, or to try grasping the possibilities of destiny, is mind-boggling to anyone's poor brain. Suffice it to say for my particular brain, the personal encounters that took place that summer in that already special place became a crossroads, a place where paths diverged in a wood and I took a road less travelled. Robert Frost took the road less travelled to become a poet. By taking my road less travelled, I was to embark on a path to an early marriage and a calling to become a pastor. Meeting Lyza Scholes was pivotal on the human level, and meeting God was pivotal on the divine. On both counts, that "has made all the difference."

"You really shouldn't wear that swimming suit around in public." I surprised myself by saying such a thing to someone I was just getting to know. We were in daily contact, since she was assigned as my waitress and I as her busboy for every meal in the dining hall. I guess that connection was a little flimsy for blurting out an observation regarding her wardrobe, or lack thereof. She cocked her pretty little head in mock shock and retorted with a quick repartee.

"And what business would that be of yours?"

I could tell by the sparkle in her eyes that she wasn't offended, but she was also not averse to taking me on. Somehow, I knew she wasn't one to be easily embarrassed. I leaned toward her ear and in a low, confidential voice said what I thought.

Cast Off Your Bow Lines

"I think your bikini is just a little too ... revealing for the eyes of your average boy around here."

I wasn't really being prudish as much as assessing the situation by my own response, which was decidedly stirring. I think I just didn't want all those other boys to be feeling the things I was feeling. She was so cute, I could hardly stand it and, yes ... sexy. Maybe it's just a matter of our particular chemistry, but my responses to her are still the same over fifty years later. That's a good and rare thing to feel for so long toward the woman of your life, but at that time it was hardly an appropriate sentiment to express. In my defense, there wasn't much but a rag of material to this so-called swimming suit, but maybe I should have kept my mouth shut.

She looked into my eyes, taking a read, and I'm not quite sure what she divined, but I think there was a certain electricity passing between us. I think she also picked up on a touch of genuine protectiveness toward her, something I was to find out later was a rare commodity in her life.

Whatever she was feeling inside, she was very good at light-hearted flirtatious banter, and she leaned back toward my ear, whispering conspiratorially, "Do you really think so?"

That's when I realized I was in over my head, and I began stammering.

"Yeah, I really think so ... at least judging by my own response ... I mean ... uh ... what I mean is I think it's just a little too ... sexy."

I think at that point she took pity on me and realized I was actually revealing something about my feelings for her, not just being judgemental about her wardrobe. Her eyes softened, and she said with sincerity, "Thanks for the tip." Then she just took off down the boardwalk, leaving me wondering.

After that episode, she didn't abandon her summer swimming wardrobe, but I was both gratified and a little disappointed to see that she used her cover up garment more regularly than she might have otherwise. I can't say for sure.

For the rest of the month after our flirtatious beginnings, we continued getting to know each other as the camp wore on. Our jobs in the dining hall afforded us many hours of unforced conversation, sitting on the boardwalk after meals, waiting for the program events to finish so we could clear our tables. I remember saying to Lyza after one particularly heartfelt conversation, "I've never told anyone that before. It's just like you read my soul." I knew I was hooked on this girl and on the depth of relationship we were forging.

Both of us had the complication of other relationships at home we would have to deal with, but our bond was irresistible, and by the end of our month of service as work crew, Lyza and I were inseparable. On the last night at Malibu, I

200

kissed her for the first time out on the golf course, and from then on through the summer and into my first year of university we became a unit to all our friends—John and Lyza. That was my Princess encounter of a lifetime. Two years later, at the tender age of nineteen, we got married, moved to Canada, and embarked upon our subsequent life adventures.

Another layer of magical experience had been laid down with Princess Louisa at the bottom of it. There were many revisits over the following years, either with our own young children during summers on our little sailboat *Fred Free* or through taking teenagers from White Rock to the Young Life camp. Sometimes there were years of hiatus, but the Princess was always there in the background, supporting us from underneath.

Many years later, our older son, Dave, and his wife, Lori, spent seven years on Young Life staff, taking teenagers to Malibu in the summers. It wouldn't be until years later, after careers and raising children were largely done, that Princess Louisa again began to loom large on our horizons. She was to become the centre of many of our coastal charter excursions.

John and Lyza at seventeen, the summer we met

Sailors and Storytellers in June, 2012

Porpoise exited the rapids with a men's group onboard, and soon we were anchored at Patrick Point and about to have dinner on the second-to-last night of the trip. I lowered the still-smoking barbequed chicken and ribs down the hatch to the crew below, where Nate Lepp sat behind the galley table in the corner of the L-shaped bench surrounded by a rowdy group of young men who hungrily awaited the latest offering. He reminded me of a mother hen with her brood. However, Nate was

Cast Off Your Bow Lines

hardly the warm, soft, and cuddly type, but rather long, lean, and strong. When he stood in the pilot house, his six-foot-seven frame kept him stooped most of the time, unless he kept his head positioned between the ceiling beams.

Despite the effort it took to stow him aboard *Porpoise*, he was a good-natured and able hand with a strong work ethic forged on the Alberta farm of his youth. He also possessed a certain charisma and routinely gathered groups of men and women for five-day cruising charters. My "First Nate," as I called him, wasn't the most likely candidate to become a sailor, but early in our charter business, he found his way onto a men's trip through my son's friends, and he fell in love with *Porpoise*, the coast, and the sailing life. No matter he was a landlubber by birth and way too tall to be "messing about in boats"—he was gut-hooked on sailing. In the ensuing years, he set about gathering his peers for sail training, and these trips eventually morphed into what he called "Sailors and Storytellers."

On these charters, after the day's adventures, one or two of the crew would share the story of their personal life journey each night after dinner. Some of those sessions got pretty emotional as the sailors shared their lives with refreshing transparency. Often the group members were all friends, but sometimes they hardly knew each other. As the week together progressed, they soon were swapping stories, sharing adventures, and forming friendships.

These kinds of trips, both men's and eventually ladies' groups, quickly became a mainstay of the shoulder season for Pacific Encounters. I found I liked hanging with young adults. I liked the generation's honesty, authenticity, and hunger for real-life experiences in the Canadian wilderness. Most of these young marrieds, professionals, or students wouldn't be able to afford my summer rates, so I charged what they could afford and made up some of the financial shortfall with higher numbers. Five or six crew made it worth it for me and doable for them, plus I got the bonus of spending time with interesting people, most of them the age of my children. I was the "old salt" with lots of stories to tell and some wisdom to dispense—just the role I liked to be in. And besides, for the cheap price, I could make them do all the heavy work. Everybody swabbed the decks, manned the sails, cleaned the fish, and washed the dishes as part of the team and crew. None of the "wait-on-you-hand-and-foot" service the better-heeled summer clientele came to expect.

As they picked over the last scraps of the ample meal and chewed the last chicken bone clean, it was apparent everybody was tired after a full day spent in Princess Louisa. The Princess produces that shock-and-awe effect on my guests every time as they stare up at the cliffs and glaciers and are gob-smacked with

Princess Encounters

her divine attributes. The young adults especially are game for the full meal deal. They do everything that can be done, from kayaking to climbing up the mountain trails to standing under the waterfalls and swimming until they become waterlogged. Of course, all of this activity is fuelled with copious amounts of food and washed down with their staple of cold beer, keeping me busy all day as chief cook (but not bottle washer).

Once they were fed and settled inside Patrick Point at the end of an exhausting day, I planted myself in a folding chair on the front deck with my own cold beer, basking in the warm satisfaction of having introduced yet another set of young souls to the magic of the Princess. I planned to head down the next morning for our final full day on Nelson Island. They would still be asleep when I caught the ebbing tide down the reaches of Jervis in the wee morning hours. By the time we arrived at our family cabin site just before noon, the guys would be raring to go again.

Unless you're born to it as I was, it's hard to get the opportunity or have the confidence to access places like Princess Louisa on our astounding British Columbia coast. As a sailing instructor, I had the ability to help these young men and women not only to experience a place like this but to teach them how to do it themselves. Sail training was an integral part of what I offered; it was fun to teach them the ways of a boat and the sea.

Canadian Recreational Yachting Association (CRYA) is a fine Canadian-based training school for the recreational boating industry, under whom I hold my credentials, and they supply me with excellent training materials. The crew members could learn how to tie knots, anchor a boat, raise and lower sails, navigate, and all the other skills needed to cruise the coast. Coastal BC is a glorious wilderness of endless islands and waterways begging to be explored. I believe the next generation is poised to become the greatest stewards of this inheritance, so I teach them to love and cherish it.

As well as the physical experience, there's a mystical part of an encounter with this land of mountains, sea, and waterfalls that calls to the heart and opens up the soul. Nate was not only exposing his friends to new dimensions of the natural world but also to dimensions of the Spirit. After becoming a qualified coastal skipper through CRYA, and after years of facilitating these trips, he furthered his credentials by becoming a qualified spiritual director. Then he became a coastal pioneer in his own right by building his own rustic cabin on Savary Island near Desolation Sound.

203

First Nate

Videographers in June, 2017

One door often opens another. Late in June, I got a call from Nate about a men's trip he wasn't even going on.

"Hey, Captain, I have a guy who wants to jump on that trip you have scheduled with Joe. He learned about *Porpoise* from seeing some pictures taken by his old chemistry teacher. You know, Barnaby, from last year's trip to Desolation Sound."

"Well sure, Joe's trip has room for one more if he wants to take another guy along."

"This new guy wants to know if you'll let him do the trip for some video footage he'll make for you."

"You mean he wants to come for free?" I was needing more income that summer and had done a lot of home video footage myself, so I wasn't too excited about that option. "Well, I'd rather have the money, but, well … sure. Tell him he can come if he at least throws in two hundred bucks for food."

Little did I know a few weeks later that a second "First Nate" had stepped aboard with Joe and all his friends. Up we went to the Princess for about the eighth time that summer. The new sailor, whose name was Levi Allen, was an affable young man who fit in immediately. As soon as we reached Jedediah Island, he began pulling a drone out of a box. The next thing I knew, the buzzing eye in the sky was skimming past me at sea level as I sat at the helm, making me feel as though the aliens had landed. Then the drone instantly shot up six hundred feet into the air.

As the footage began to come in on the monitor, I realized I had a professional on board. Never had *Porpoise's* attributes been shown to such advantage and in such a setting, especially later, when under full sail, making passage in a brisk northwesterly up Jervis Inlet. I couldn't pay for the kind of publicity those

Princess Encounters

videos would provide. My new young man was a full-on videographer with many thousands of Instagram followers and his own production company, Leftcoast Media House. He also ran sought-after Wilderness Adventure Film workshops. I told Levi he could forget about chipping in money for groceries on that trip.

Later the following year, some young web builders on board proposed to trade me a website upgrade for a sailing trip. They said, "We love *Porpoise* and these cruises, but we think your website needs an upgrade. It's a bit old school." They used Levi's videos as well as some others from previous trips and built a new website that made Pacific Encounters look really cool. From then on, *I* was cool. In ensuing years, Levi started to bring aspiring videographers from all over the world to stage wilderness videography workshops in the context of a five-day sailing trip to Princess Louisa and beyond. Needless to say, the Princess's features are now being shown to best advantage all over the world, and I continue to get to hang with the young and with-it types.

Levi Allen of Leftcoast Media House

Princess Contemplations

One day as I sat reading a book in a folding chair on *Porpoise*'s deck, and Lyza kayaked in the glassy waters at the head of Louisa, I began to meditate upon it all. We were anchored in our special spot beneath ancient maples and evergreens, whose trunks were covered with lichen and whose heavy branches were draped with long strands of trailing moss, evidence of the weather that usually prevailed in this rainforest. Ferns and other flora were even rooted high up in the crotches of these trees. Because the anchorage was so deep, we were tucked in close to shore and stern-tied to an old snag. I had to crane my neck upward to see the welcome blue sky above the looming cliffs that towered over us. It was a place that

Cast Off Your Bow Lines

could produce an impending sense of doom in a person given to claustrophobia. The masses of rock above were so ominously weighty that one could feel crushed under their mere presence. From a distance out in the inlet, *Porpoise* appeared under these great cliff faces as a tiny toy boat. It was a place to feel significant and insignificant simultaneously.

I put down Hemingway's *The Sun Also Rises*, in which the following Bible verse was quoted—"*One generation passeth away; and another generation cometh: but the earth abideth forever*" (Ecclesiastes 1:4, KJV). I considered the stupendous granite face above Chatterbox Falls, where my favourite cascade sprang out of nowhere from the very top of the six thousand-foot ridge to free-fall a thousand feet and explode upon a ledge. From there it dropped and sluiced from one crag to another until it finally slipped along the lower faces and disappeared into the forest, where it passed chattering unseen through rocky beds and pellucid pools to finally emerge frothing among the boulders along the water's edge. There were so many waterfalls coming down it was impossible to know which stream came out where.

How long had that cataract been falling? I wondered. It never stops. It was doing the same thing in my great grandfather's time and, of course, way before that, throughout the years of native habitation and back into the dim mists of time, flowing, crashing, and forever falling into the same inlet.

"*The earth abideth forever …*" I thought of how the eyes of six generations of my own blood had gazed upon the same unchanging scene. My great grandfather, grandfather, father, myself, my children, and grandchildren had now all meditated upon the same monolithic scene. Six generations. Wouldn't seven, the biblical number of perfection, just complete the tale? I didn't want to think yet about the prospect of becoming a great grandfather, but I was aware that if my eldest granddaughter, Irelynn, had gotten married and had a child at the age Lyza and I did, she would already have a baby by now.

How privileged we were to have spent a whole lifetime exposed to these beauties year after year. The layers of memory overlapped in my mind until the time periods became indistinguishable. It made me think of how apt the title was of M. Wylie Blanchet's coastal classic, *The Curve of Time,* to describe the interconnected years on this coast. In the dog-eared copy of her great book that I always kept close at hand on the boat, she describes her experiences with her children and the places and peoples of the coast from the Gulf Islands to the far reaches north of Vancouver Island. Her summer journeys mostly took place in the thirties (when my grandfather cruised), but in her telling, time becomes irrelevant and the images registering in the reader's imagination are like viewing slides

Princess Encounters

in an old-fashioned carousel tray. It's no longer important which year it is—it's a continual stream of consciousness embracing the generations. There is only you, your boat, your people, and the moment you're in as you drift along under the same unchanging backdrop—the mountains, sea, and sky.

A kaleidoscope of images, real and imagined, shuttled through my mind one after another until they blended into one timeless river of consciousness. I could hear the voices of my ancestors calling from amid the trees and almost see their forms walking upon the ridges of rock or swimming and laughing in the wild waters. I could sense the Creator's presence in the landscape, but I also felt Him permeating the chain of lives and experiences recorded in this place.

As my reverie continued, I couldn't help but ponder and deeply appreciate my rich life upon these waters. Here I was, fifty-some years since my fourteen-year-old cutthroat experience, still looking for lakes, fishing for trout, waiting for the treasure of the day to be revealed in some new guise. I was still exploring, dreaming, discovering—waiting for the next gem to uncover itself. I might catch a glimpse of the glory of it in the latest mountainscape as it reflected the last rays of the setting sun. It might be in the form of a merganser preening its rust-coloured head feathers against the white and grey of its back. It could suddenly break out of the waters off the bow as a huge humpback whale breaching for the fun of it. The treasure could be found in the radiant smile of a child from Ethiopia or in the repose of a grandparent from the Canadian Prairies drinking in the verdant views. It could be found in the satisfactory sighs of a group of friends pushing back from the galley table after a seafood feast, or in their lively conversation recounting their very special day. How amazing it still was to me to be the skipper of such a sailing ship after thirty-five years of an entirely different kind of career. What a gift from God to be called back to the roots of my childhood and be able to share this world with others. It wasn't what I'd ever expected or thought possible, but you've read our story now and have seen how it was.

The next morning Lyza and I trekked down the long reaches of Jervis Inlet and along the north shore of Nelson Island until our little cove hove into view. We always look forward to some downtime in our own special place and to a gathering of the clan. This particular summer, everyone's plans had serendipitously come together at the same time and we could already see the green hull of *Ariel* tied in her spot.

At that time, our daughter, Julie, her husband, Chris, and their two sons, Kai and Jude, were taking their turn cruising on the little sailboat. Matt, our son, sat at the mouth of the cove on the bench he had crafted himself out of stones and shells from the beach. He yelled, "Ahoy, *Porpoise*!" in his booming voice. We could see the colourful forms of various grandchildren scampering over the rocky shoreline. When they caught sight of us coming around the bend, they danced and shouted to us their excited greeting. Dave, our older son, and his wife, Lori, were lounging on the cabin's deck, cold beers in hand. Their four children—Irelynn, Finn, Tate, and Sailee—were part of our welcoming committee on the beach. Up on top of the little island in the mouth of the cove, Gigi and her daughter, Marieani, waved hello, big grins on their faces.

    This would be a spontaneous gathering of the clan with lots of sharing of the summer's adventures, lots of swimming and snorkeling, campfires and s'mores, and lots of love and laughter. I swung *Porpoise* around in the mouth of the cove to drop anchor and back into the narrow slot, where we cast two stern lines to hold us taut between the adjacent rocky shores. It was a tricky manoeuvre perfected by much practise, and soon we were snug and secure in our little redoubt.

Porpoise secured in the mouth of our cove

Princess Encounters

PPSS

A Cabin by the Sea
(a family prophetic poem)

When the sun shines on the fir trees, it's there that I would be,
There beside the forest, there beside the sea.

When the sea fog shrouds the landscape and hides the dripping trees,
I'd love to be among them and hear the sounding seas.

I'd love to have a cabin where I could take my ease,
Snug above the rocky beach and safe among the trees.

I'd love to feel the salt spray and walk along the shore
And hear the scuttling shellfish and the ocean's solemn roar.

I'd love, Oh! how I'd love again, to look upon the sea,
To see and feel it all again in quiet reverie.

The seagull in the sky above, the shellfish on the shore,
Oh! How I'd love the sea again and the ocean's solemn roar.

I'd love to see the sunshine and the water's emerald blue
And see again the rocky cliffs stand out in varied hue.

And then again, I'd love to see the shrouding, blinding fog,
The mists that hang o'er the islands and drip from tree and log.

I'd love again to see those logs which the tide makes rise and fall,
Which line the shore for miles and miles from out the forest tall.

Oh! I'd love to see it all again, the great, mysterious sea,
And gaze o'er its wide, wide waters in solemn reverie.

I'd like to look out Westward, out where the red sun dips
Into the far flung ocean plowed by the stately ships.

209

Cast Off Your Bow Lines

I'd love to see the fir trees and hear them softly croon
And watch the big wide ocean 'neath the whiteness of the moon.

Oh! I'd surely love to be there in a cabin by the sea,
Where the song of eternal harmony would come wafting in to me.

                Charles Clarke 1931

The cabins by the sea and a rowboat full of grandchildren

Another Book by John Clarke

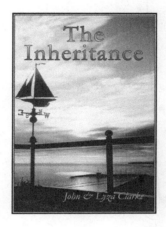

When John and Lyza took a sabbatical twenty years ago, he wrote his own *Summa Theologica*. Unlike Aquinas's, John's is not a weighty tome but rather another burst of storytelling, one of his favourite avocations.

Nevertheless, in spite of the storytelling, or more accurately because of it, *The Inheritance* is actually pretty deep. By taking the reader on another sailing trip, building a cabin in the wilderness, exploring his ancestral roots in Vinalhaven, Maine, and living by the sea in Maui, John makes an insightful parallel between the Exodus story, his own life journey, and the events that befall them on their sabbatical. He will make the case there are seven stages in the spiritual journey and ten tests of life that all of us will face. The story becomes the lesson, and the lesson becomes the story as the insight about our eternal inheritance is intriguingly illustrated.

For more information about Pacific Encounters Sailing Adventures,
see the website

www.pacificencounters.com